T0360672

SCIENCE, MEDICINE, AND THE AIMS OF INQUIRY

After its unparalleled rise and expansion over the past century, medicine is increasingly criticized both as a science and as a clinical practice for lacking scientific rigor, for contributing to overmedicalization, and for failing to offer patient-centered care. This criticism highlights serious challenges which indicate that the scope and societal role of medicine are likely to be altered in the twenty-first century. Somogy Varga's ground-breaking book offers a new perspective on the challenges, showing that they converge on fundamental philosophical questions about the nature and aim of medicine. Addressing these questions, Varga presents a philosophical examination of the norms and values constitutive of medicine and offers new perspectives on how to address the challenges that the criticism raises. His book will offer valuable input for rethinking the agenda of medical research, health care delivery, and the education of health care personnel.

SOMOGY VARGA is Professor of Philosophy at Aarhus University, Director of the Center for Philosophy and the Health Sciences, and Senior Research Associate in the Faculty of Humanities at the University of Johannesburg. He is the author of *Scaffolded Minds* (2019), *Naturalism, Interpretation, and Mental Disorder* (2015), and *Authenticity as an Ethical Ideal* (2011).

SCIENCE, MEDICINE, AND THE AIMS OF INQUIRY

A Philosophical Analysis

SOMOGY VARGA

Aarhus University

Shaftesbury Road, Cambridge CB2 8EA, United Kingdom

One Liberty Plaza, 20th Floor, New York, NY 10006, USA

477 Williamstown Road, Port Melbourne, VIC 3207, Australia

314–321, 3rd Floor, Plot 3, Splendor Forum, Jasola District Centre, New Delhi – 110025, India

103 Penang Road, #05–06/07, Visioncrest Commercial, Singapore 238467

Cambridge University Press is part of Cambridge University Press & Assessment, a department of the University of Cambridge.

We share the University's mission to contribute to society through the pursuit of education, learning and research at the highest international levels of excellence.

www.cambridge.org
Information on this title: www.cambridge.org/9781009450010

DOI: 10.1017/9781009449977

First published 2024

A catalogue record for this publication is available from the British Library

A Cataloging-in-Publication data record for this book is available from the Library of Congress

ISBN 978-1-009-45001-0 Hardback

Contents

Preface and Acknowledgments

The impetus for this book was born out of an increasing awareness of what appears to be intensifying criticism of modern medicine. After what many regard as the "golden age" of medicine in the twentieth century, prominent figures now maintain that medicine has entered "the age of disappointment." Medicine is increasingly criticized as a science and as a clinical practice for lacking proper scientific rigor, contributing to overmedicalization, and failing to offer compassionate, patient-centered care. The criticism highlights serious challenges to medicine which indicate that its scope and societal role are fated to be altered in the twenty-first century. Such a critical threshold provides fertile ground for addressing fundamental questions about medicine, and this book takes on such a task.

The book illustrates that the criticism and the challenges it conveys converge on fundamental, philosophical questions: What is the (scientific) nature of medicine? What is the aim of medicine? The book offers a systematic philosophical examination of these questions, persuaded that such an undertaking carries the potential to assist in the approaching deliberation about the future of medicine. It defends three theses about the nature and aim of medicine (the *Systematicity Thesis*, the *Understanding Thesis*, and the *Autonomy Thesis*) that illuminate norms and values constitutive of medicine. The theses enable us to assume the *Moderate Position* with respect to the challenges, which offers a better comprehension of the problems, points toward possible solutions, and helps to rethink the proper boundaries of medicine and the appropriate use of medical means.

I have benefited enormously from exchanges of various lengths with Martin Marchmann Andersen, Alexander Bird, Jens-Christian Bjerring, Alex Broadbent, Anke Bueter, Nancy Cartwright, Remy Debes, Anna Paldam Folker, Jonathan Fuller, David Miguel Gray, Klemens Kappel, Andrew James Latham, Sigurd Lauridsen, Rune Munck Aabenhus, Lennart Nordenfelt, Jennifer Radden, Heidi Samuelson, Jacob Stegenga, Asbjørn Steglich-Petersen, Fredrik Svenaeus, Bettina Utzelmann, and

Vincent Varga. The various interactions with these individuals were pivotal in refining the core claims advanced in this book as well as amplifying their clarity. A special thanks to Heidi Samuelson for remarkable feedback and exceptional copyediting and to Vincent Varga, my eldest son, for his assistance in reviewing and discussing some of the historical content contained in this work. Lastly, I extend my gratitude to two anonymous reviewers whose constructive feedback significantly contributed to enhancing the quality of the manuscript.

Despite the challenging circumstances surrounding the COVID-19 pandemic, during which significant portions of this book were written, I was fortunate to present parts of this work to various audiences. Thanks to audiences at Durham University, University of Pittsburgh, University of California San Diego, University of Memphis, Södertörn University, University of Toronto, McGill University, and my graduate students at Aarhus University. I am grateful for the opportunity to engage with these diverse groups in discussions about various aspects of the book. Some of the material in the book is significantly based on articles that have been published in journals. Thanks to the publishers for the permission to use material from the following pieces: "The Aim of Medicine. Sanocentricity and the Autonomy Thesis," *Pacific Philosophical Quarterly* (2023) 104: 720–745; "The criticism of Medicine at the End of its 'Golden Age'," *Theoretical Medicine and Bioethics* (2022) 43(5–6): 401–419; "The Continuity of Inquiry and Normative Philosophy of Science," *Metaphilosophy* (2021) 52 (5): 655–667; "Understanding in Medicine," *Erkenntnis* (2023) doi.org/10.1007/s10670-023-00665-8; "Medicine as Science. Systematicity and Demarcation," *Synthese* (2021). 199(1–2): 3783–3804.

I am grateful to Hilary Gaskin at Cambridge University Press for support and encouragement from the start. The completion of the research for this project was made possible by a generous research grant awarded by the Carlsberg Foundation.

Introduction
Medicine at the Crossroads

I.1 The Road to Medicine's "Golden Age"

Among our most universal and intimidating human experiences, the confrontation with disease occupies a prominent rank. Throughout history, a plethora of varied forms of folk and professional healing practices have aimed to fight disease, to enhance the ability to cope with it, and to render it meaningful. But during the course of the nineteenth century, the emergence of scientific medicine quickly changed the relatively pluralistic landscape of healing practices. Major discoveries such as the cell at the center of pathological changes and the "germ theory of disease" dealt a deathblow to theories that had dominated Western medicine since the ancient Greeks (e.g., the humoral theory of disease). Advances in physics, chemistry, and biology converged to form a basis for the field of medicine, which rose from a craft based on observation accumulated at the patient bedside to the level of a respectable branch of science. Since then, apart from scientific and technological advances, two mutually reinforcing tendencies have fueled the unparalleled rise and expansion of medicine.

The first is the *socialization of medicine*, which refers to the increasingly organized allocation of public funds to more or less centralized health services. Policymakers discovered the potential of medicine. In times of peace, the efficient functioning of complex industrial economies required a population that was both literate and healthy. In times of war, substantial injections of public funds into centralized health care services helped to sustain civilian morale and keep soldiers in the field. The provision of health services through subsidized medical facilities (e.g., dispensaries and hospitals) and national insurance plans became a way to attain political stability, to moderate the menace of sickness among poor people, and to control dangerous environmental conditions caused by massive industrialization. By the mid-twentieth century, citizens of financially recovering Western European states had access to a variety of state-supported medical schemes. In the US, a national health program did not develop, but in spite of an ideological commitment to private medicine, the government carried a growing share of health care through the Armed Services, the Veterans Administration, and the Public Health Service (Porter 2002).

The second tendency is what we could call the *medicalization of the social realm*, which refers to a development whereby medicine expands its reach into an increasing number of private and public areas of life. An increasing number of previously nonmedical conditions came to be defined as medical conditions, which required diagnosis, prevention, or treatment (Broom and Woodward 1996; Conrad, Mackie, and Mehrotra 2010). Social movements functioned as promoters for medicalization, helping change the understanding of behaviors formerly defined as deviant (i.e., immoral, sinful, or criminal) into disease symptoms. A number of conditions like alcoholism, psychopathy, eating disorders, sex addiction, and learning disabilities turned from badness to sickness. Alcoholism is a good example of a case in which a new understanding of alcoholism as a disease was chiefly accomplished by the efforts of a social movement (Alcoholics Anonymous), while the medical establishment initially held back. As the reach of medicine widened, emerging statistical knowledge about the distribution of disease and its relationship to other variables (e.g., class, education, housing, diet) gave rise to population-level measures targeting the seemingly healthy (e.g., screening, testing, prenatal care) and to appointing doctors as factory inspectors, medical officers, analysts, and forensic specialists to help implement policies and laws (e.g., food and drug control, workplace safety, sanitation).

Socialization and medicalization played substantial roles in the development of medical science in the twentieth century (Porter 1997; 2002; Le Fanu 2012). Medicine transformed from a small-scale practice into an immense global industry, and from a craft to something that many have regarded as an emblem of scientific progress living up to the ideals of the Enlightenment, overcoming ignorance, and superstition for the benefit of mankind. In particular, the mid-twentieth century is often portrayed as the "golden age of medicine," an era characterized by scientific and therapeutic advances and high levels of public esteem bestowed upon the medical profession. To mention a few achievements, the main dangers to human life before World War II that were responsible for the deaths of millions of people (e.g., septicemia, tuberculosis, pneumonia, tetanus, polio, syphilis, meningitis) became treatable or preventable by vaccination by the 1980s.[1]

[1] When the World Health Organization (WHO) declared the global eradication of smallpox in 1980, medicine won a victory over a disease that is estimated to have killed 500 million people during just the last century of its existence.

The discovery of insulin as a treatment for type 1 diabetes marked a key breakthrough, and so did the invention of high-tech tools that enable early diagnosis (e.g., cytogenetic, biochemical, and molecular testing) and formerly inconceivable forms of surgery (e.g., organ transplants, laser surgery).

It is important to point out that the talk of a "golden age" sometimes underestimates the effect of factors such as better living standards on improvements in health (McKeown 1976; for a discussion, see Bynum 2008; Kaplan and Milstein 2019) and that "revolutionary narratives" about advances sometimes interfere with more nuanced and critical analyses of therapeutic successes and contributions to longevity (Farmer, Basilico, and Messac 2016). But while it is important to keep in mind that most of the gains in the first half of the century are associated with improved nutrition, sanitation, housing, and public health measures, medical advances (new drugs, devices, and procedures) have been a significant source of increases in longevity since World War II (Cutler, Deaton, and Lleras-Muney 2006b; Fuchs 2010).[2] This is true even if these successes coexist with numerous areas in which cures continue to elude medicine's reach (e.g., cures for influenza, fibromyalgia, cancer, Parkinson's, schizophrenia), or in which available interventions are of limited effect (e.g., statins, type 2 diabetes drugs, selective serotonin reuptake inhibitors [SSRIs]).

I.2 The Criticism of Medicine and the "Age of Disappointment"

Echoing themes from earlier criticisms of medicine as well as adding new aspects, the beginning of the twenty-first century has witnessed the emergence of a critical movement that advocates the reevaluation of medicine's efficiency and societal role. Even though the social determinants of health are now more widely acknowledged (Aronowitz 2019), leading medical professionals, epidemiologists, and historians, as well as some voices among the general public, express doubt that the status medicine enjoys in contemporary Western societies is justified (see, e.g., Porter 2002;

[2] Estimating the impact of medical advances in improving life expectancy is a highly complex issue and it is difficult to control for nonmedical factors. Perhaps Fuchs's conclusion, which attributes to medical advances the primary role in increasing life expectancy, is too optimistic, and improvements in life expectancy are probably not best pursued via investments in medical services (Kaplan and Milstein 2019). However, the more modest view that medicine has made a significant contribution is hard to deny. Studies focusing on a smaller number of conditions offer a clearer picture: half of all health improvements between 1960 and 2000 are due to medical care (Cutler, Rosen, and Vijan 2006a; Cutler et al. 2006b) and a considerable part of the reduction in infant mortality can be interpreted as resulting from improved neonatal medical care for low birth-weight infants (Cutler 2004).

Stegenga 2018; Broadbent 2019). For example, in a publication in *The Lancet*, prominent gastroenterologist Seamus O'Mahony (2019a) argues that we have entered the "age of disappointment," characterized by declining trust in medicine and growing criticism of it.

Three forms of criticism stand out, each highlighting challenges to contemporary medicine. First, *skeptics*, who rank among the most prominent and respected physicians and epidemiologists, maintain that confidence in the effectiveness of many medical interventions ought to be low (see Stegenga 2018). Numerous medical interventions are unsuccessful, and many others do not fare considerably better than a placebo. Moreover, we often see evidence suggesting that an intervention is effective even when it is not, in part because the institutional structure producing medical research is biased in favor of positive evidence and against reporting negative findings.

Second, critics maintain that *overmedicalization* occurs, meaning that medical resources are improperly used to address political, social, and personal problems and turn these problems into pathological conditions (Parens 2013). The expansion of the category of what demands medical intervention is often driven by predominantly social judgments about what is considered appropriate (in terms of body, behavior, personality, etc.). This contributes to the explosion of the costs of medical treatment, and leads to overtreatment (Scott 2006; Conrad 2007).

Third, critics target what might be called *objectification* in medical care. Increasingly technologically mediated interaction contributes to discounting the personal experience of illness and the psychological and social dimensions of ailments (Cassell 2004; Marcum 2012; Topol 2019). It predisposes seeing the body of the patient as a system made up of interacting and separately operating parts, increasing the likelihood of medical professionals forgetting that they are engaged with persons in vulnerable states (Engelhardt and Jotterand 2008). These issues lead to an increasing dissatisfaction in patients, which may be one of the reasons for the growing popularity of complementary and alternative medicine (Astin 1998; Bivins 2010).

Taken together, the charge is that medical science is less trustworthy than generally thought (skepticism), medical means are used to address nonmedical problems (overmedicalization), and medical care is inadequate (objectification). The criticism is thus comprehensive because it targets medicine both as medical science and medical practice, claiming that medicine has diverted from its course such that its aim fails to be realized in the current institutional settings.

I.3 The Aims of the Book

The sheer scope and depth of the criticism and the problems it highlights suggest that medicine has reached a critical threshold and indicate that medicine's scope and role in society is fated to be altered in the twenty-first century. This situation provides a fertile ground for addressing fundamental, philosophical questions about medicine. In particular, the different strands of criticism seem to converge on fundamental questions, namely about (a) the nature of medicine and (b) the aim of medicine, while also implicating central concepts in medicine such as health and disease. First, whether medicine can be justifiably accused of failing as a science (skepticism) will depend on what its nature is, that is, to what extent it can be adequately described as science. Second, whether the charge of overmedicalization is warranted will depend on what the aim of medicine is. If medicine is aimed not merely at fighting disease, but at enhancing well-being in the widest sense, then the charge might not be justified. Third, whether the charge of objectification is vindicated will depend on what the aim of medical care is. If it is merely the removal and prevention of disease, then the charge might not be justified.

If these considerations are on the right track, then we may presume that a systematic philosophical examination of these fundamental questions carries the potential to assist in the approaching deliberation about the future of medicine as a science and clinical practice. Aspiring to assist such a deliberation, this book pursues three main goals. It offers:

(1) an account of the *nature* of medicine

(2) an account of the *aim* of medicine

(3) a *Moderate Position* based on these two accounts that rethinks the challenges to medicine and outlines possible solutions.

Much of the current literature operates with more or less implicit assumptions about the nature and aim of medicine. With respect to the question about the nature of medicine, one influential view is that medicine is something other than science, as it merely applies science and does not pursue knowledge for its own sake (see, e.g., Pellegrino 1998; Miller 2014; Miller and Miller 2014). With respect to the question about the aim of medicine, the standard view is that the aim of medicine is to cure diseases, or at least to deliver proper care by using the arsenal of available medical interventions. However, as we shall see in the course of this investigation, neither of these answers, nor recent alternatives, is satisfactory, which obstructs productive debates about medicine. Instead, this book proposes and defends more precise formulations of three broad theses about the nature and the aim of science and medicine:

Systematicity Thesis:	Medicine is science, that is, systematic inquiry.
Understanding Thesis:	Scientific inquiry in medicine aims at understanding.
Autonomy Thesis:	The primary purpose of understanding in medicine is to promote health, pursued to the extent that it serves or is at least consistent with the final aim of promoting autonomy.

To reach its first two goals, the book gathers support for these theses. Doing so involves illuminating norms and values that are constitutive of medicine, which, when suitably explicated, offer valuable impulses for dealing with the challenges that critics draw attention to. Helping to reach the third goal of the book, the theses will allow us to assume the *Moderate Position* with respect to the challenges, which provides a better comprehension of them, points toward possible solutions, and helps rethink the proper boundaries of medicine and the appropriate use of medical resources.

To reach its objectives, the book outlines a particular way of philosophically engaging science and medicine that guides the investigation. It develops an approach, best described as a *normative philosophy of medicine*, which focuses, for example, not only on what medicine is, but also on what it should be, and not only on how medical knowledge is deployed, but how it should be deployed. The approach operates with *three levels of analysis* and shows that the current criticism and challenges to medicine require addressing basic questions on all three levels. The approach can be located at the intersection of two different philosophical approaches to medicine. One largely pursues analytic aspirations, clarifying metaphysical and epistemological issues in order to analyze theoretical and practical aspects of medicine. The other largely pursues normative aspirations, aiming to comprehend ethical issues in health care and apply ethical reasoning to assist decision-making. Neither of these is entirely suitable for the tasks of this book, as the separation of analytic and normative considerations would risk overlooking how fact and value are often inescapably joined in the realm of medicine.

I.4 The Structure of the Book

Chapter 1 lays the groundwork and contributes to comprehending normative issues in medicine by offering an analysis of three prominent forms of criticism that target contemporary medicine. First, the chapter explores *skeptical criticism*, which maintains that except for a few "magic bullets," confidence in the effectiveness of medical interventions ought to be low (see Stegenga 2018). Second, the chapter surveys the criticism of *overmedicalization*, defined, as opposed to the neutral term "medicalization," as the improper usage of medical resources to address political, social, and personal problems. Engaging critics (e.g., Moynihan and Cassels 2005; Conrad 2007; Le Fanu 2012; Parens 2013), five reasons are presented for thinking that overmedicalization is problematic. Third, the chapter explores the charge of *objectification*, which raises vital questions about medical care (e.g., Cassell 2004; Haque and Waytz 2012; Capozza 2016; Topol 2019). The chapter clarifies this criticism and explores technological mediation and deindividualization in health care environments as contributing factors.

The last part of the chapter draws on work on criticism (e.g., Popper 2000), arguing that two features unite the three predominant forms of criticism. The first feature is that the criticism is *social* in the sense that its object is a social practice and not merely the actions of individuals engaged in the practice (see Haslanger 2018). The second feature is that the criticism is *internal* in the sense that the standards of evaluation employed are internal to the practices criticized and not external and independently justified. It is argued that the three forms of criticism build on the implicit assumption that medicine fails to meet *its own* internal standards: it has diverted from its course such that its aims are not adequately promoted in current institutional settings. But then, the different strands of criticism seem to converge on more fundamental questions about (a) the aim of medicine, (b) the nature of medicine, and (c) the key concepts of health and disease. The vast majority of the chapters in this book (Chapters 3–7) are predominantly dedicated to addressing these fundamental questions.

In order to achieve these aims, Chapter 2 presents and defends a particular type of philosophical engagement with medicine that guides this book: *the normative approach*. In a critical dialogue with existing work on normativity in the philosophy of science (e.g., Sober 2008; Kitcher 2011; Kaiser 2019), the chapter outlines a normative approach to the philosophy of science that distinguishes between *three levels of analysis* (i.e., aims, nature, and key concepts), corresponding to the types of questions that current challenges to medicine raise. This grounds a particularly attractive *normative approach to philosophy of medicine* that considers philosophy of medicine as a proper subdiscipline of philosophy of science.

The approach deserves the label "normative" for several reasons. It uncovers norms linked to the aims, nature, and key concepts of medicine, assesses to what extent they are actually fulfilled in practice, and offers corrections based on these findings. Moreover, it allows for evaluating the merits of the current criticisms of medicine, and parts of the chapter are dedicated to showing how its three levels of analysis can contribute to addressing the criticism and challenges to which Chapter 1 drew attention. But, it is important to highlight that – consistent with what Chapter 1 said about internal criticism – the approach is normative in a particular manner: it is a second-order philosophical inquiry that is *continuous* with normative elements that are already more or less explicitly present in medical science and clinical medicine. This aspect, call this the Continuity View, breaks with an influential tradition in the philosophy of medicine, which sometimes openly advocates, and sometimes implicitly assumes, that philosophy is discontinuous with science and that philosophical work on medicine is therefore "detached from the method and content of medicine" (Pellegrino 1986, 13; 2001). The chapter offers a defense of the Continuity View against objections that could be launched by proponents of a traditional view in the philosophy of medicine like Edmund Pellegrino. It is shown that due to its scope and the three levels of analysis it highlights, the normative approach displays important advantages compared to traditional accounts and is particularly well positioned to help reach the goals of the investigation in this book.

In order to approach particular questions about the nature of medicine and the extent to which it is a genuinely scientific enterprise, Chapter 3 addresses the general question about the nature of scientific activity with special attention to medicine. While one influential view is that medicine is something other than science, as it merely applies science and does not pursue knowledge for its own sake (see, e.g., Pellegrino 1998; Miller 2014; Miller and Miller 2014), one main task is to defend the *Systematicity Thesis*, according to which medicine is science, that is, systematic inquiry.

The chapter starts by consulting the literature on the "demarcation" problem in the philosophy of science. It is argued that the failure of well-known approaches should not lead us to abandon the issue, but rather to pose the demarcation question in a different manner and proceed without entertaining essentialist expectations and hence ahistorical or discipline-independent necessary and sufficient conditions. Science is best seen as a family resemblance concept, and the most promising way to consider the sciences as united is not through some intrinsic property, but a relational property that only admits differences of degrees to nonscientific inquiries. The *Deflated Approach* adopted in this chapter is based on Paul Hoyningen-Huene's (2013) account of systematicity as a necessary condition for science. It is shown that medicine (i.e., medical science *and* clinical medicine) meets the requirement for systematicity. Of course, the fact that medicine fulfills a necessary criterion for science does not establish that it is one, but as it displays systematicity on all the considered dimensions and is more systematic than its everyday counterpart, we have good reasons to think of medicine as science. In the last part of the chapter, it is shown that the *Systematicity Thesis* is able to differentiate medicine from activities widely recognized as pseudoscience. In a critical dialogue with recent work (e.g., Oreskes 2019), the chapter shows that homeopathy does not exhibit the type of synchronic and diachronic systematicity that characterizes scientific endeavors and it therefore remains susceptible to a variety of biases. Systematicity helps generate reasoning and inquiry that produce reliable knowledge and understanding.

The defense of the *Systematicity Thesis* helps clarify the nature of medicine in terms of systematic, scientific inquiry. But what is the aim of scientific inquiries in medicine? Focusing on the *epistemic* aim of inquiry, Chapter 4 seeks to make a critical step toward answering this question by focusing on medical science, which, as described in Chapter 3, encompasses clinical as well as medical laboratory research, and only counts as properly *medical* if it displays a *practical orientation,* that is, if it is ultimately motivated by contributing to the maintenance of health and the diagnosis, prevention, and treatment of disease. The main thesis of the chapter, the *Understanding Thesis*, holds that inquiry aims at *understanding*, while the question of what special kind of understanding is at stake in medicine is the topic of subsequent chapters. Drawing on recent debates in epistemology (e.g., Kvanvig 2003; Pritchard 2010; Grimm 2014) and in a critical interchange with prominent work in the philosophy of science (Kitcher 2001; 2008; 2011; Bird 2007; 2019a; 2019b; Douglas 2009; Potochnik 2017), the arguments presented in favor of the *Understanding Thesis* break with an influential view that, due to its practical orientation, inquiry in medicine differs *in kind* from scientific inquiries, leading them to the conclusion that "medicine is not, and cannot be, a science" (Munson 1981, 189; Pellegrino 1998; Miller and Miller 2014). The chapter shows that the success of this argument depends on faulty assumptions about the aims of scientific inquiry. It is argued that the practical orientation of inquiry in medicine does not render it different *in kind* from scientific inquiries, and does not prevent it from being a science. However, there are important differences in degree, which make a difference for what counts as progress. Finally, the *Understanding Thesis* has some implications for thinking about responsibilities in scientific inquiry, which are clarified by *extending systematicity* to include considerations about the choice of an inquiry.

The starting point of the subsequent chapters ensues from a number of points made in previous chapters. If the *Systematicity Thesis* and the *Understanding Thesis* are correct, then we may derive the broad suggestion that (i) the aim of medicine is to understand pathological conditions, which (ii) serves the final objective to contribute to the endeavor of supporting human agency. After all, if the epistemic interest in understanding is motivated by practical interests, and if pathological conditions are in general detrimental to human agency, then it makes sense to assume that the goal of understanding pathological conditions is to be able to intervene (i.e., cure, treat, prevent them) in a way that promotes our abilities as agents. However, both (i) and (ii) deserve more detailed consideration, as much will depend on *what exactly* the character of understanding is in medicine and *how exactly* medicine contributes to supporting human agency. For this reason, Chapter 5 focuses on (i) while Chapter 6 deals with (ii).

Chapter 5 starts out with exploring a simple suggestion that has roots in the *Understanding Thesis,* and its main task is to shed light on the specific kind of understanding that medicine has as its aim. Taking into consideration work by Alex Broadbent (2019), current debates on the epistemology of understanding (e.g., Kvanvig 2009; Grimm 2012; Khalifa 2017), and recent scholarship on the aims of inquiry (e.g., Kelp 2021), the chapter first describes in more detail what it means to understand something, distinguishes types of understanding, and considers the history of scurvy to explore what understanding a disease involves in the context of medicine. The main hypothesis here is that *objectual understanding* of a disease (i.e., biomedical understanding) requires grasping a mechanistic explanation of that disease.

To see how causal and constitutive relationships are comprehended in the sciences, the chapter draws on an influential account of causation (Woodward 2003; 2010; 2015) and on work on mechanistic explanations in the biological sciences and neuroscience (Thagard 2003; 2005; Craver 2007; Nervi 2010; Kaplan and Craver 2011; Darrason 2018). However, alluding to debates on methodological principles in the humanities and social sciences, it is argued that biomedical understanding is necessary but not sufficient for understanding in a clinical context (i.e., clinical understanding). Rather, clinical understanding combines *biomedical understanding* of a *disease* with *personal understanding* of an *illness*. In some cases, personal understanding is extended, necessitating the adoption of a particular second-personal stance and using cognitive resources *in addition* to those involved in biomedical understanding. The attempt to support this hypothesis will include revisiting the distinction between "understanding" and "explanation" familiar from debates concerning methodological principles in the humanities and social sciences. While reflection on the everyday use of "knowledge" and "understanding" will offer guidance, it will not be sufficient for tackling substantial questions in the context of scientific inquiry. Thus, consistent with what was said about explication and conceptual engineering in Chapter 2, the task is not merely to analyze ordinary concepts, but to "engineer" appropriate concepts such that they assist with advancing the inquiry.

Chapter 6 also receives guiding impetus from the *Systematicity Thesis* and the *Understanding Thesis*, and it explicates how exactly understanding in medicine contributes to supporting human agency. The chapter starts by examining an initially plausible proposal according to which medicine is pathocentric (e.g., Pellegrino 2001; McAndrew 2019; Hershenov 2020), aiming to restore the health of individuals by curing or treating disease. Discussing and rejecting this opening proposal as well as competing ideas, the chapter presents and defends the *Autonomy Thesis*, which holds that medicine is not pathocentric, but aims to promote health with the final aim to enhance autonomy (understood as including competency and authenticity conditions; see Christman 2009). Drawing on accounts in which health is more than the absence of disease (e.g., Venkatapuram 2013; Nordenfelt 2017), the chapter defends and adopts a "positive" notion of health and clarifies its relations to other concepts such as well-being and autonomy. It draws on the normative approach outlined in Chapter 1 and on recent work on "conceptual engineering" (e.g., Chalmers 2020) to offer a pluralist perspective on some difficulties surrounding the concept of health. It closes by considering and defusing the objection that the *Autonomy Thesis* is overly permissive and allows many highly controversial procedures (e.g., forms of elective cosmetic surgery, prescribing steroids for athletes) as legitimate parts of medicine.

Two methodological considerations guide the chapter. First, the inquiry is limited to "mainstream medicine" (i.e., scientific Western medicine) that – at least on some level of abstraction – is sufficiently universal in spite of variation in local features of institutions and practices (see Broadbent 2019, ch. 1). Second, the question about the aim of medicine is unearthed in tandem with a closely connected matter that concerns the "internal morality of medicine," that is, the moral norms and values that govern the practice of medicine (e.g., Brody and Miller 1998; Pellegrino 2001; Ben-Moshe 2019). These norms generate prima facie moral obligations on medical professionals independently of general morality and offer a normative backdrop against which inappropriate use of medical understanding can be identified.

While no narrowly confined aim will be able to capture the full complexity of medicine, the task here is to offer an account that is sufficiently broad to help address the challenges, yet sufficiently narrow to pass the requirements of philosophical rigor. Even though the chapter mainly focuses on clinical medicine, it is argued that aims pursued by other branches of medicine that acquire population-level data and biological knowledge of health and illness are fused with those of clinical medicine. Thus, the aim of medicine is explicated on a general level, such that it applies to clinical medicine but will also have implications for medical science, social medicine, and preventive medicine.

Chapter 7 continues our reflections on the aim of medicine, but is dedicated to exploring and critically engaging contemporary accounts from the literature. While the accounts by Edmund Pellegrino (2001) and Alex Broadbent (2019) each identify an overarching aim of medicine, the chapter will also consider four "list approaches" that each offer a catalogue of aims, including the Hastings Center Report (Callahan et al. 1996) and lists produced by Howard Brody and Franklin G. Miller (1998), Bengt Brülde (2001), and Christopher Boorse (2016). This is a rather considerable amount of material to explore in a single chapter; however, the aim of the reconstruction is not to do justice to many of their details, but to focus on examining to what extent they are able to overcome or bypass the challenges faced when defending the *Autonomy Thesis*. Subjecting these contemporary views to critical scrutiny is not merely an essentially adversarial procedure, but is also a means to assist framing the proposal presented in the previous chapter. By inspecting the most relevant aspects of these accounts in light of the challenges considered in Chapter 6, the chapter also provides further reinforcement for the *Autonomy Thesis* by considering paths that the proposed account chose not to take.

Chapter 8 returns to the challenges to medicine that have motivated the investigation in this book. The chapter is dedicated to showing how the account defended in this book can help address the three strands of criticism (skepticism, overmedicalization, and objectification) and the challenges to medicine they draw attention to. As the challenges converged on fundamental questions about medicine, and as the three theses defended in previous chapters (i.e., the *Systematicity Thesis,* the *Understanding Thesis,* and the *Autonomy Thesis*) have made progress on these issues, the chapter rethinks the challenges in light of the findings. Taken together, the three theses allow us to take up what we could call the *Moderate Position* that situates itself between more radical views. With respect to each of the challenges, the *Moderate Position* offers a better comprehension of the relevant problems, points toward possible solutions, and adds clarity to our thinking about the proper boundaries of medicine and the appropriate use of medical means.

Rethinking the challenges conveyed by skepticism, it is argued that they can be understood as violating the norms of scientific inquiry outlined by the *Systematicity Thesis.* Taking up a more *Moderate Position* than skeptics, the chapter shows that increasing systematicity in simple and extended senses offers ways to address the challenge of skepticism. Bringing into play the *Understanding Thesis,* it is argued that increasing systematicity would also require reconsidering resource allocation in medical research to prioritize certain research goals.

Rethinking overmedicalization in light of the *Autonomy Thesis* leads to a *Moderate Position* on the proper boundaries of medicine. Against a number of critics, it is argued that the medicalization of a condition does not amount to overmedicalization as long as the condition is harmful (i.e., causes or significantly increases the risk of suffering, harm, or death), and medicine offers an adequate understanding of it in the sense elaborated in Chapter 5. If these conditions are fulfilled, then the medicalization of a condition is consistent with the aim of medicine, regardless of whether medicine can offer effective treatment for it.

Finally, rethinking objectification in light of the *Autonomy Thesis* shows how objectification can hinder the pursuit of the aim of medicine by obstructing personal understanding (as outlined in Chapter 5), which thwarts systematic inquiry and can have a detrimental effect on the aim of promoting health and autonomy (as described in Chapter 6). At the same time, the *Moderate Position* highlights the essential function of standardization and technological advances for attaining systematic biomedical understanding, but stresses that medicine will fail to harvest the full benefits of these unless it implements measures that counteract the diminished personal understanding (and objectification) that they can contribute to.

I.5 Final Remarks

This book is motivated by current challenges that provide a fertile ground for addressing fundamental questions about medicine. It illustrates how the recent criticism of medicine converges on fundamental philosophical questions, outlines a specific approach (i.e., the *normative philosophy of medicine*), defends a novel account of the aim of medicine (i.e., the *Autonomy Thesis*) and the nature of medicine (i.e., the *Systematicity Thesis* and *Understanding Thesis*), and shows how these offer a new perspective on the challenges to medicine (i.e., the *Moderate Position*). The book tackles such questions hoping to assist an informed deliberation about medicine, to generate a constructive impact on rethinking its future trajectory, and to inspire further work at the intersection of philosophy and medicine.

The book contributes to the quickly growing field of philosophy of medicine, which thus far features relatively few book-length works. It discusses and supplements two important books on philosophy of medicine, namely Jacob Stegenga's *Medical Nihilism* (2018) and Alex Broadbent's *Philosophy of Medicine* (2019). Stegenga's book focuses on defending the view that we should have little confidence in the effectiveness of medical interventions, and it offers a defense of a hybrid theory of disease. This book has different aims, operates with a broader notion of what constitutes a medical intervention, and offers an account of positive health associated with the aim of medicine. Broadbent's book defends a theory of the nature of medicine that clarifies its purpose and meaning (i.e., medical cosmopolitanism), and it operates with a broad notion of what constitutes medicine, which also allows for exploring debates between competing medical traditions. While Broadbent's book focuses on what unites different traditions, this book focuses on making sense of a particular tradition in a "bottom up" fashion. Also, another unique feature of this book is that it proceeds by integrating new developments in epistemology and philosophy of science (e.g., on understanding, progress, and inquiry) as well as debates on the internal morality of medicine.

Because the book incorporates new developments in epistemology and philosophy of science, it covers a relatively large amount of philosophical terrain and will mostly appeal to philosophers. Nonetheless, the book also hopes to speak to a philosophically informed readership that is interested in medicine, its place in society, and the moral and epistemic norms that guide medical research and medical care in various settings. The account presented in the book might be interesting to philosophically informed health professionals seeking new input for prioritizing research goals, for reflecting on the allocation of health care goods, or for pondering the relationship between alternative, traditional, and mainstream medicine. In addition, clarifying these issues might assist decisions in cases in which curing disease, promoting individual health, and promoting public health are at odds with one another.

CHAPTER 1

Challenges to Medicine at the End of Its "Golden Age"

1.1 How Medicine Became the Patient

During the nineteenth century, advances in physics, chemistry, and biology converged to form the basis for scientific medicine. Since then, medicine has achieved a historically unparalleled global dominance, grown into a global industry, and changed the previously pluralistic landscape of healing practices throughout the world. The expansion reached its zenith during the second half of the twentieth century, which historians and medical professionals often portray as the "golden age of medicine" (Porter 2002; Kernahan 2012; O'Mahony 2019a; 2019b),[1] characterized by high levels of prestige and confidence in medical institutions and in the efficacy of medical interventions.

Whether such confidence is based on measurable therapeutic successes and contributions to longevity is, however, contested, and "revolutionary narratives" about advances sometimes interfere with more nuanced analyses (Farmer et al. 2016). Already in the nineteenth century, pathologist Rudolf Virchow maintained that "the improvement of medicine may eventually prolong human life, but the improvement of social conditions can achieve this result more rapidly and more successfully" (quoted from DeWalt and Pincus 2003). A century later, physician Thomas McKeown argued that the reduction of mortality observed during the twentieth century is largely attributable not to medicine, but to better nutrition, housing, and public health measures (McKeown 1976).[2] Writing in a time

[1] The exact temporal boundaries of the "golden age" are not drawn consistently in the literature. Some maintain that the mid part of the twentieth century constitutes the golden age, sometime after World War II (Kernahan 2012), while some associate it with the "conquest" of epidemic infectious disease (Brandt and Gardner 2020).

[2] Some of McKeown's most forceful claims were based on studying mortality decline in England and Wales. Since then, researchers have pointed to similar examples during the mid-twentieth century (China 1949–79; Cuba 1959–79) where medicine has played only a minor role in mortality decline compared to improvements in housing, sanitation, and education (Farmer et al. 2016).

characterized by radical criticisms of societal institutions, McKeown's work became part of a critical movement that advocated the reevaluation of medicine's efficiency and societal role.

On the more radical side of this movement, some argued not only that sanitation, nutrition, and housing were more important determinants of health than medicine, but also that medicine has become an institution of social control (Zola 1972) and a threat to health (Illich 1974). Ivan Illich's 1974 paper "Medical Nemesis" in *The Lancet*, followed by his bestselling attack on modern medicine with the same title distinguished three types of iatrogenesis: clinical (i.e., direct harm by treatment), social (i.e., medicalization of life problems), and cultural (i.e., the loss of traditional ways of dealing with suffering). However, his radical indictment of medicine as "institutional hubris" (Illich 1974, 922) and his calls for "deprofessionalisation of medicine" (Illich 1974, 921) were dismissed by many medical professionals. His criticism was polemic, radical (i.e., maintaining that medicine probably did more harm than good), selective (e.g., downplaying successful aspects of medicine in relief and rehabilitation), driven by a more general criticism of modernity, and, importantly for our purposes, it came from outside of medicine.

Today, almost five decades after the publication of "Medical Nemesis," medicine is increasingly subject to various forms of criticism that raise themes familiar from Illich's work. The criticism is much more comprehensive, nuanced, and comes from inside medicine, that is, from leading medical professionals, which makes it harder to ignore. For example, in a publication in *The Lancet* (2019a) and a book with the evocative title *Can Medicine Be Cured?* (2019b), prominent gastroenterologist Seamus O'Mahony notes that since entering medicine toward the end of its "golden age," he has witnessed decline and corruption in medical research and medical practice. He raises questions about three aspects of medicine.

First, O'Mahony maintains that "medical research . . . has itself become a patient," increasingly scrutinized by metaresearchers. Second, he argues that "medicine has extended its dominion over nearly every aspect of human life," herding "entire populations – through screening, awareness raising, disease mongering, and preventive prescribing – into patienthood" (O'Mahony 2019a, 1798–9; 2019b, 25–6). Third, he laments having witnessed "the public's disenchantment with medicine," as expressed in patient reports of their experiences in various health care settings. Patients have become, as O'Mahony (2019b, 330) puts it, "a problem to be processed by the hospital's conveyor belt; it is hardly surprising that they often feel that nobody seems to be in charge, or cares about them as individuals."

The criticism by leading medical professionals like O'Mahony is worthy of further investigation, especially since the criticism and other challenges facing medicine (e.g., an aging population, explosion of costs) seem to indicate that medicine's scope and role in society is fated to be altered in the twenty-first century. At this critical threshold, providing a firm grasp of dominant forms of criticism, explicating the norms they appeal to, and exploring the problems they draw attention to can assist an informed deliberation about the future of medicine.

To contribute to achieving this task, the chapter proceeds in three steps. First, it distinguishes three sorts of criticism that O'Mahony's work touches on, but that can be found expressed in much greater detail elsewhere in the literature.[3] The criticism raises questions about medical research (skepticism), the use of medical means to address nonmedical problems (overmedicalization), and about features of medical care (objectification). Second, upon distinguishing forms of criticism and the nature of the norms they appeal to, it is argued that the criticism of medicine is *social*, *internal*, and appeals to *constitutive norms* of medicine. Third, it is argued that the criticism converges on more fundamental questions about (a) the aim of medicine, (b) the nature of medicine, and, less directly, (c) key concepts in medicine. Addressing these questions will not only help assess the criticism, but also contribute to a deliberation about the role of medicine in the twenty-first century.

1.2 Skepticism

When O'Mahony maintains that medical research "has become the patient," he touches on a growing skepticism about whether the status and confidence that medicine has enjoyed in contemporary Western societies is justified. The roots of skepticism go back to the 1960s, when the prestige of the medical establishment suffered from catastrophic effects of new drugs,[4] as well as from the recognition that environmental hygiene, improved nutrition, and better living standards have contributed more than clinical medicine to guaranteeing longer lifespans.

[3] While the comprehensive analysis of forms of criticism does not rest on O'Mahony's observations, his work is useful to mention, not only because it briefly introduces topics from the perspective of a physician, but also because it is a prominent example of relatively fierce criticism from a medical professional published in a leading medical journal.

[4] Perhaps, most notably, thalidomide, which caused deformities in more than 10,000 newborns and provoked firmer regulations for drug licensing.

In the contemporary landscape, we may distinguish two types of skepticism (see also Stegenga 2018; Broadbent 2019). *Historical skepticism* argues that mainstream medicine only merits its status since the emergence of modern clinical trials and since it acquired a genuine capacity to extend life during the mid-twentieth century.[5] Prior to this stage, despite all the progress in science, over the course of two thousand years medicine only achieved a few reasonably effective interventions (e.g., quinine for malaria, orange and lemon juice for scurvy, opium for pain relief, colchicum to treat gout, amyl nitrate to dilate arteries, herbal preparations as purgatives), and doctors knew that many of their interventions were ineffective (Porter 1997; 2002; Wootton 2006). Improvement had been achieved by discontinuing certain procedures (e.g., bloodletting) and introducing new procedures (e.g., hand washing), while other improvements (e.g., the retreat of diseases like diphtheria, typhoid, and tuberculosis) were attributable to better diet, housing, and working conditions. Even the striking victory against disease due to the introduction of a smallpox vaccination "came not through 'science' but through embracing popular medical folklore" (Porter 1997, 11).

Worse, some argue that prior to the twentieth century, medicine might have done more harm than good, in part because it long held on to interventions based on humoral theory, which were ineffective or even detrimental to health.[6] For some two thousand years, alongside purging and vomiting, the principal therapy was bloodletting (phlebotomy or venesection), which weakened and sometimes even killed patients. Without the concept of infectious disease and persuaded that no two illnesses are identical, effectiveness could not be measured, and the commitment to this tradition often outweighed any contrary evidence.[7] The emergence of larger hospitals in the eighteenth century, often seen as a sign of great progress, in

[5] Indeed, evidence-based medicine – a movement stressing that clinical decisions ought to be made on the basis of the best available evidence of effectiveness – is in part motivated by recognizing that the history of medicine is dominated by harmful or ineffective interventions.

[6] The humoral theory of disease in some general form remained popular among physicians until the mid-nineteenth century. However, there were also prominent exceptions. For example, William Harvey's discovery in the seventeenth century that blood circulates in the body via a closed system of vessels contradicted what the *humoral theory* predicted about the motion of blood (Wootton 2006, 47–8 and 95–6).

[7] A good example is the history of scurvy, a condition caused by vitamin C deficiency, of which Chapter 5 offers a detailed discussion. The important point here is that although sailors already knew about the effectiveness of lemon juice as a prophylactic in the early seventeenth century, physicians trained in humoral pathology resisted the idea even after some initial clinical studies confirmed what the sailors reported. Instead, they remained confident that the correct diagnosis was "humoral imbalance" and that the condition called for bloodletting and vomiting induced by salt water (Wootton 2006, 162–3).

many cases actually made medicine more dangerous. For example, while mothers and infants had previously been relatively safe in the care of informally trained midwives, nineteenth-century hospitals significantly increased the risk of death, because doctors inadvertently spread infections from one patient to the other on their instruments and hands.

Contemporary skepticism draws attention to present-day challenges and is promoted by some prominent and respected physicians and epidemiologists. In the extremely influential article "Why Most Published Research Findings Are False," published in the journal *PLoS Medicine*, John P. A. Ioannidis explored the reliability of published medical research findings and concluded that the majority of published research claims are false (Ioannidis 2005; see also 2016).[8] In similar ways, prompted in part by escalating health care costs and the growing preparedness to render medicine more evidence based, a growing amount of meta-research casts doubt on the efficacy of some widely used treatments, identifying factors that can influence the choice of topic, study design, and methodology in ways that potentially undermine the validity of published research findings. Building on this line of research, Jacob Stegenga (2018) argues that it is difficult to vindicate our confidence in the efficiency of contemporary treatments to eliminate the symptoms and underlying causes of disease. In fact, Stegenga (2018, 11) concludes that except for a few "magic bullets," we ought to have low confidence in the effectiveness of interventions.

Stegenga (2018) formulates the argument by using Bayes's Theorem, which calculates the probability of a hypothesis (H) provided evidence (E) that appears to support H. The probability of H given E, $P(H|E)$, depends on three other probabilities: (i) the prior probability of H being true, regardless of E (i.e., $P(H)$); (ii) the probability of the evidence given H (i.e., $P(E|H)$); and (iii) the prior probability of E, irrespective of H (i.e., $P(E)$). The resulting equation, $P(H|E) = P(H) \times P(E|H) / P(E)$*, states that the probability of H given E is equal to the prior probability of H, multiplied by the probability of E given the hypothesis, divided by the prior probability of E.[9]

[8] Many have since criticized some of the radical statements in Ioannidis's work, arguing that the chosen model incorrectly lowers the evidential value of studies (see, e.g., Goodman and Greenland 2007). However, what is more important for our purposes is that the critics tend to agree with Ioannidis's general points about the challenges in medical research and that the problems with different forms of bias are more severe than generally assumed.

[9] $P(H|E)$ is low if (i) $P(H)$ is low (i.e., if it is unlikely, regardless of E, that H is true), (ii) $P(E|H)$ is low (i.e., the observed E is not very probable given H), and (iii) $P(E)$ is high (i.e., it is very likely that E would be observed regardless of whether H is true).

How does this support the skeptical thesis that we ought to have low confidence in the effectiveness of interventions? For P(H|E) to be low (i.e., the posterior probability of the medical intervention is effective, given evidence that appears to support its effectiveness) three conditions have to be met.

(1) P(H) is low (i.e., the prior probability that a particular medical intervention is effective is low).

(2) P(E|H) is low (i.e., the evidence observed is improbable given the hypothesis that the intervention is effective).

(3) P(E) is high (i.e., the prior probability of observing evidence that supports the intervention is high, regardless of whether it is actually effective).

Stegenga offers support for the thesis that these three conditions are met in current research, and the main points may be summarized as follows. In support of (1), one can give an inductive argument from the fact that most of the medical interventions tested prove unsuccessful. Drug companies test a large number of interventions that fail and never make it to the market. But even among those that pass the tests and reach the consumer, a significant number are later restricted or entirely withdrawn.[10] It is certainly possible that highly successful "magic bullets" (e.g., antibiotics, vaccines, insulin for diabetic treatment) that target a highly specific cause of disease in an effective manner (without many side effects) will be discovered. But there are reasons to remain skeptical about the chances of such discoveries. First, magic bullets are "low-hanging fruit," which means that most of them have probably been discovered already. Second, it is very difficult to devise an intervention that is both highly specific and effectively targets diseases with complex and poorly understood underlying causal mechanisms. The current tools for intervening, like various forms of chemotherapy, are often rather crude and nonspecific.

In support of (2), Stegenga stresses that in many cases interventions are little better than placebo, that effect sizes in trials tend to be low, and that studies frequently reach discordant results (Stegenga 2018, 171–5). Good examples of particularly low effect sizes in widely prescribed medications include antidepressants and cholesterol-lowering drugs (statins). The best available evidence suggests that they have minimal positive effects. While selective serotonin reuptake inhibitors (SSRIs) only do slightly better than

[10] Examples include isotretinoin, rosiglitazone, valdecoxib, fenfluramine, sibutramine, cerivastatin, and nefazodone.

a placebo at managing depression (Kirsch 2019), statins lower cholesterol levels, but fail to clearly decrease mortality in asymptomatic patients: in order to avoid a single death from any cause, physicians have to prescribe them to about 244 people with no history of heart disease for five years (see Redberg and Katz 2016). Moreover, the evidence for effectiveness is uncertain in many cases, with some studies suggesting positive effects, while others suggesting no effects or negative effects.

In support of (3), it is to be expected to find evidence indicating that an intervention is effective *even if it is not*, in part because the institutional structure of medical research is biased in favor of positive evidence. Evidential standards (e.g., meta-analyses and systematic reviews, hierarchy of evidence, randomized controlled trials) do not eliminate problems with the malleability of research methods. Meta-analysis involves subjective judgments about inclusion criteria, the weight given to studies, and the correct interpretation of the results, such that two groups of researchers analyzing the same evidence can report different conclusions. The inter-rater and inter-tool reliability for assessing the quality of evidence is not very high, which means that such studies may not be able to effectively identify biases. In addition, the structure of medical science might incentivize exploiting the malleability of the methods to produce evidence of positive effects, especially in cases in which trials are conducted by the companies who manufacture the products being tested. Potentially aggravating problems of malleability, pharmaceutical companies and scientists have a vested interest in reporting positive effects, while there is a bias against reporting negative findings and no incentive to replicate findings.[11]

Overall, bearing in mind the factors that support (1)–(3), the limitations of professional standards (e.g., peer reviewers can evaluate the quality of the submitted studies, but these might consist of a biased sample of the total evidence), and the limits of regulatory oversight, the upshot is this: even after taking into account seemingly solid evidence in favor of the effectiveness of an intervention, we ought still to assign only a low probability to the claim that it is effective. The skeptical conclusion in Stegenga's work is based on an inference to the best explanation that uses

[11] For an example, consider the case of a drug for type 2 diabetes, rosiglitazone (Stegenga 2018, 148). A lawsuit required the manufacturer GlaxoSmithKline to disclose the entire dataset accumulated from forty-two trials. It turned out that only seven trials had published their results, all of which suggested that the drug was effective. The drug was approved by the FDA in 1999, but a meta-analysis based on the data from all forty-two trials found that it increased the risk of heart attack by 43 percent (Nissen 2010). While on the market, the drug is estimated to have caused more than 80,000 heart attacks.

numerous examples and is supported by the identification of methodological, social, and financial factors. It is consistent with the fact that we generally underestimate the role of nonmedical interventions like changes in hygiene and nutrition in improving health and the role of medicine in adverse health outcomes.[12] Accepting the conclusion does not require denying the possibility of genuine medical breakthroughs (e.g., genetic engineering leading to highly effective interventions), but it supports a skeptical attitude toward claims about them. One important limitation that we will return to in Chapter 8 is that the assessment of the effectiveness of medical interventions is based on a narrow notion of what constitutes "medical," construed essentially as using pharmaceuticals to target diseases.

1.3 Overmedicalization

The second issue that O'Mahony mentions is linked to the fact that, parallel to the ascent in the standing of the medical profession, a growing number of issues and conditions came to be portrayed and comprehended in medical terms, including pregnancy, obesity, alcoholism, lack of success in education, and drug addiction. During the 1970s, the term "medicalization" was coined to describe such processes by which conditions previously considered as nonmedical are increasingly defined and treated as medical problems (typically as illness, disorder, or disease) and handled by medical professionals. For example, saying that pregnancy has been medicalized means that pregnancy is now seen as a potential disruption to health that requires expert medical care and risk management.

As such, medicalization is a value neutral, descriptive term designating cases in which medical means are used for conditions hitherto considered as outside the medical realm. For instance, medicalization occurred when a set of problems known as shell shock was redescribed as the symptoms of the medical condition post-traumatic stress disorder (PTSD), or when alcoholism was transformed from a moral to a primarily medical problem. The identification of a previously overlooked disease can also be seen as the result of medicalization, and so can efficient birth control (Parens 2013).

[12] In a widely discussed paper, Makary and Daniel (2016) have calculated that every year more than 250,000 preventable deaths occur from medical mistakes in the US alone. While this number is likely inflated due to methodological issues (see Shojania and Dixon-Woods 2017), other studies find that approximately 3.5–4.5 percent of hospital deaths are due to preventable medical error.

In contrast, *overmedicalization* involves the improper use of medical resources. Of course, improper use as such is not sufficient for overmedicalization: physicians assisting in torture arguably put to use medical resources in an improper fashion without being involved in overmedicalization. Instead, overmedicalization refers to the improper use of medical resources to address political, social, and personal problems, often replacing established practices that traditionally addressed them. It can occur in two ways. A condition can be medicalized with or without *pathologization*, that is, attaining the label of a pathological condition (Sholl 2017). Overmedicalization that does not involve pathologization describes a change toward comprehending various types of medical interventions as justified with respect to a condition. Critics typically speak about overmedicalization in this sense with respect to pregnancy, fertility, and death. Overmedicalization that involves pathologization describes how certain conditions that enter the medical jurisdiction become labelled as pathological (e.g., alcoholism, epilepsy). Critics typically argue that a category error occurs that turns life problems and normal human variations into pathological conditions (Parens 2013). For example, while individuals living in social isolation due to being severely shy and socially awkward were traditionally not considered as suffering from a medical condition, they are today increasingly diagnosed with mental disorders like social phobia or social anxiety disorder, which imply some difference in kind from "normal shyness."

Critics argue that overmedicalization has a number of potentially severe consequences. First, by expanding the category of what demands medical action, overmedicalization increases the number of people deemed to be in need of medical intervention by many millions and contributes to the explosion of the costs of medical treatment. In the case of social phobia or social anxiety disorder, at any given time, almost 5 percent of the US population meets the diagnostic criteria. For some, this shows that overmedicalization is driven by medical industries that stand to earn massive profits by classifying as pathological conditions that were previously perceived as variations of normal states.[13] Such practices of "disease-mongering" aim to increase the number of people who can be diagnosed by relaxing diagnostic criteria, by constructing more or less bogus disease categories, or by transforming risk factors or precursors to disease into diseases. For example, the decision to lower the diagnostic threshold for high cholesterol

[13] The FDA gave permission to advertise certain SSRIs like paroxetine as a drug for social phobia; SmithKline launched an effective campaign ad with the slogan, "Imagine being allergic to people."

was surrounded by controversy, not merely for clinical reasons, but also because the vast majority of the experts on the panel that revised the relevant guidelines had financial ties to pharmaceutical companies that manufactured cholesterol-lowering drugs (Moynihan and Cassels 2005).

Some of the structural characteristics of regulating the industry offer financial incentives for disease-mongering. For example, a company can hold on to the protection of a patent in case a new use for the product is developed. Eli Lilly supported defining a new disease called premenstrual dysphoric disorder (PMDD), rebranded fluoxetine as Sarafem, and received FDA approval for promoting Sarafem for PMDD. This secured the extension of their patent on fluoxetine, at a time when many disagreed that PMDD was a genuine disease. The FDA approval in fact came before the APA decided to recognize PMDD as a distinct psychiatric condition (see, e.g., Mintzes 2006).

Second, the worry is that overmedicalization in some cases does not reflect clinical observations or findings, but predominantly social judgments about what is considered to be appropriate behavior (Scott 2006; Conrad 2007). More precisely, overmedicalization might be taken to reflect disapprobation of behavior that is perceived as failing to conform to dominant values in contemporary culture. In the case of shyness, the relevant dominant values are those attached to being self-confident, talkative, assertive, and comfortable with self-presentation.

Third, overmedicalization changes the focus of problem-solving to individual-level medical interventions and away from the political and social structures that generate conditions under which being severely shy is increasingly a debilitating problem. This obstructs the emergence of genuine public deliberation that might lead to rethinking whether the relevant dominant values in contemporary culture – such as the value of capacity to perform with ease in the social realm – should be resisted. Such deliberation might lead us to revise entrenched ideas about the acceptable norms of navigating social situations in a way that would allow recognizing a larger natural variation in social skills and behavior.

Fourth, overmedicalization appears to be causally implicated in an increase in the number of healthy people who are seriously concerned about their health. In a development that seems puzzling in light of gains in lifespan and health, people increasingly see their lives as acutely threatened by real but trivial risks or sometimes by downright fictional hazards (e.g., cell phones, low radiation) (Le Fanu 2012). It is highly probable that the explosion of conditions and risk factors that are now classified as pathological has contributed to this development.

Fifth, overmedicalization may lead to overdiagnosis and overtreatment, for example, in cases in which physicians accurately diagnose a patient as having the pathophysiological basis P of a disease D, where P would never have led to symptoms of disease D and would not have interfered with the patient's life. This case becomes one of overtreatment if the patient in question is treated for D by intervening on P. Overtreatment in this case does not directly result from lowering the thresholds for D, but from deploying more precise ways of detection in very early stages. For example, some argue that screening programs for prostate cancer lead to overdiagnosis and overtreatment (Loeb et al. 2014). The point is that a large number of patients diagnosed with prostate cancer might receive unnecessary treatment, as they would not develop symptoms if left untreated.

1.4 Objectification

Finally, O'Mahony identifies a different and growing problem with respect to medical care. Dissatisfaction with mainstream medicine among patients as well as practitioners has grown during the last decades, amplified by the implementation of new managerial strategies and cost-capping initiatives (in welfare states) and by growing suspicion that medicine is excessively driven by profit (e.g., in the US). Critics argue that mainstream medicine fails to offer empathetic care driven by patient need. Patients seek not only scientifically based management of their conditions, but also what is often described as a "humane" care that also addresses the existential or psychological aspects of those ailments. They want to be relieved and cured, but they also seek explanations of their predicaments, a sense of wholeness, and control (Porter 1997, 51). Patients increasingly complain that such needs are not met and that the care they receive is often "objectified" or "dehumanized." Without being able to do justice to the full complexity of the phenomenon, some clarification can be achieved by briefly examining factors like *technological mediation* and *deindividualization* in health care environments that critics link to objectification.

First, the advances in therapeutic and diagnostic devices have contributed to the emergence of technologically mediated management that suppresses dimensions of care that would address the psychological and social dimensions of ailments (Blumer and Meyer 2006; Marcum 2012). The emphasis on this type of management and its increased dependence on sophisticated technology stimulates the tendency to sideline the patient's illness experience from the clinical consultation. It predisposes physicians toward perceiving the body of the patient as a system constituted by cooperating and separately operating parts, and such a focus contributes to perceiving the patient's individuality, subjective experience, and personal narrative as something that risks obfuscating direct access to the disease. The patient as a person is at risk of disappearing in the encounter, eroding the conditions for an intimate relationship with a medical professional that many patients associate with earlier stages of medical practice. Critics argue that with this development, medicine has lost something crucial. As Cassell (2004, ii) puts it, medical doctors are now "less skilled at what were once thought to be the basic skills of doctors – discovering the history of an illness through questioning and physical examination, and working toward healing the whole person" (see also Weatherall 1996, 17).

Second, health care environments tend to deindividualize both patients and physicians, which probably contributes to the experience of receiving objectifying care, sometimes also described as "dehumanizing." In a mutually reinforcing process, the deindividualized appearance of the patients (e.g., wearing uniform coats and gowns) might make them appear less as individual agents that require empathy, while the deindividualized appearance of the physicians (e.g., wearing uniform white coats) might mask their individual responsibility toward patients. The nature of these environments might also contribute to practices that increase objectification. For example, patients are sometimes labeled in terms of their illnesses ("diabetic" instead of "a person with diabetes") or referred to by acronyms or by the body part being operatively intervened on, both of which collapse the distance between the person and the disease (see, e.g., Haslam et al. 2007; Todres, Galvin, and Holloway 2009; Haque and Waytz 2012). Such practices increase the likelihood of medical professionals forgetting that they are engaged with people who are in vulnerable states, who grant them access to highly private aspects of their life, and whose trust they need in order to be able to care for them (Engelhardt and Jotterand 2008). Highly specialized health care that focuses entirely on the disease often translates illness experiences into several different diagnoses in a way that does not render their predicament transparent and meaningful to the patient, leading to experiences of objectification. As a patient puts it, "you do not feel human, but . . . as an object on a conveyor belt, no one really cares. They have decided, medical science has determined, that's the way it is" (Berglund et al. 2012). Such reproaches do not target human error in the work of physicians or nurses, but systemic problems and institutional culture.

Of course, voicing these concerns does not require denial of the numerous benefits associated with using technologically sophisticated devices or the benefits of focusing more narrowly on less than the whole human being in diagnosis and intervention.[14] For example, pharmacological treatments of mental disorders may necessitate switching from the language of subjective symptoms to that of biochemical processes, even if it may lead to patients feeling objectified. Also, in certain scenarios, there might be some benefits associated with not focusing on the patient as a fully social being. The training of physicians encourages effective regulation of empathy, which dampens emotional responses that result from perceiving others suffering and can help physicians deal with stress (Di Bernardo et al. 2011). Many medical procedures involve inflicting pain on the patient, and it is likely that such procedures could be efficient without significantly reducing the distress that comes with causing pain.

The experience of unmet needs during medical care may be one of the reasons for the growing popularity and prominence of "complementary and alternative medicine" (CAM). This describes a broad range of health practices that historically originate outside of conventional or mainstream medicine (e.g., acupuncture, herbal remedies, naturopathy, homeopathy, and chiropractic), which position themselves in relation to mainstream medicine, but differ in their attitude to it. In spite of efforts by medical authorities to keep in check the proliferation of CAM services and products, the National Health Interview Survey from 2007 reveals about 40 percent of US residents use at least one CAM health practice, and the number of visits to providers of CAM outnumbers the visits to primary care physicians practicing mainstream medicine (Barnes, Bloom, and Nahin 2007).

While this popularity is in part explained by increased economic wealth that stimulates the consumption of "health products" (e.g., vitamins, plant extracts, etc.), it is also linked to the perception of mainstream medicine as objectifying, and sometimes also authoritative and bureaucratic. Although the motives are not entirely clear, what patients describe as a lack of bedside manner and an objectifying environment in mainstream medicine is one of the reasons for the popularity of alternative medicine (Astin 1998). As Bivins (2010, 36–7) puts it, "the rigors of biomedicine from the patients' perspective – the degree to which it was impersonal, driven by and constructed around the needs of the laboratory and technology . . ., and disease- rather than patient-focused – provoked many to accuse both

[14] We should also note that some authors who explicitly recognize that "the practice of medicine has been progressively dehumanized" insist that technologically sophisticated devices also offer the solution by freeing up time for "human-to-human bonding" (Topol 2019, 491–2).

the medical system and its practitioners of arrogance, insensitivity, and greed." Along such lines, some movements entirely reject mainstream medicine and denounce it as a part of an elitist "conspiracy against the laity" (Porter 2002, 45).

1.5 The Character of the Criticism

Taken together, the three forms of criticism are comprehensive and highlight substantial challenges: medicine is scientifically less rigorous and trustworthy than generally thought (skepticism), medical resources are improperly used to address nonmedical problems (overmedicalization), and the care received violates expectations (objectification). The criticism thus targets medicine as both a medical science and a medical practice, and it constitutes a powerful assembly of forces that will contribute to transforming medicine in the twenty-first century. At the same time, the criticism is *nuanced* in the sense that it simultaneously recognizes that medicine is facing different challenges than just a century ago. Critics are well aware that increased longevity due to advances brings to the fore a range of chronic diseases that are much more difficult to treat. Also, they are aware that medical professionals increasingly encounter individuals with composite medical and social needs (e.g., related to homelessness and substance abuse), and it would be unrealistic to expect that professionals with medically defined roles would be able to meet these needs.

The following sections further explicate the criticism and the challenges to medicine it highlights. One important step toward completing this task is to unearth the specific normative character that the different forms of criticism share. Focusing on the nature of the standards of evaluation that they deploy can assist a better understanding of the criticism but also provide clues as to how to deal with the challenges to which they point. Before we start, two notes on the choice of terms are in order.

First, a note on what "criticism" means. In this case, as well as in general, criticism aims to raise awareness of a problem and contribute to changing the state of the target, which can be some state of affairs in the world, or (e.g., historical skepticism) the stance that we take toward it. Importantly, while change can be effectuated in a number of ways (e.g., using monetary incentives, threats, manipulation), criticism aims to change things by offering reasons. For this, besides appealing to certain observed facts, it has to appeal to some *norm* that purports to provide a reason and thereby *justify* change (Kauppinen 2002). Norms specify standards that can be met or fail to be met; they prohibit and permit courses of action, but they also implicitly structure the space of possibilities of action (Jaeggi 2018, ch. 3). Norms are linked to values, on the one hand (e.g., courage is a general value, norms define what is courageous behavior in a situation), and to reasons, on the other. A justification can be suitably demanded for why norms should be met, but in many cases they are profoundly implicit, such that it would not make sense to demand one.[15]

Second, some distinguish between "criticism" and "critique" and take the former to refer to something less elaborated and directed toward persons and the latter to be a more developed consideration upon a subject. However, this distinction is ambiguous and not used systematically in the literature. For example, in his discussion of criticism in science and philosophy, Popper (2000) consistently speaks of "criticism," even though the way he uses the term fits the definition of "critique." For this reason, we will use "criticism" in a broad sense, which includes instances of "critique."

1.5.1 Ways to Criticize Social Practices

Criticism can target individuals, actions, states of affairs, or, as in our case, a *social practice.*[16] Roughly, a social practice is a collective activity that involves an arrangement of norms, and it functions, as Sally Haslanger (2018, 237) puts it, "in the primary instance, to coordinate our behavior around resources." Practices are defined as "offices and positions with their rights and duties" (Rawls 1971, 55), including procedures for determining admissible and nonadmissible violations.[17] Practices can be conceived in terms of norm-conforming behavior, but it is essential that the norms and rules inherit their purpose and point from the *aim* of the practice and the good at which it is directed (MacIntyre 2007). These constitutive aims (e.g., the law aims at justice, education at developing children's abilities, medicine at health) provide criteria for evaluating the behavior of participants. The practice may require institutions to serve its aim by norm enforcement, organization, and funding, and these norms may be changed

[15] It is customary to distinguish types of norms, rules (e.g., games), prescriptions (e.g., legislator and legal norms), and directives (e.g., technical instructions) (Wright 1963).

[16] The criticism only targets the action of individuals indirectly, to the extent that these are constitutive parts of the practice.

[17] It is not clear, however, that practices can be said to be governed by rules. Drawing on Wittgenstein's work on rule-following, some have argued that rules are more or less adequate representations of aspects of practices that are prior to the rules. Rules, as Wittgenstein puts it, cannot keep participants in practices "on the rails" of the practice. Being able to comprehend what it is to follow a rule might require a prior conception of practice.

in a way that advances the aim of the practice, without transforming it into something different. Of course, the relationship between institutions and practices is more complicated. An institution is not itself structured by the aim and norms of the practice it organizes, but in terms of practice-external goods like status, money, and power (MacIntyre 2007, 194). Because institutions have a tendency to separate from the practice they sustain, the pursuit of two kinds of goods constitutes a source of potential conflict. For MacIntyre, without virtues (e.g., truthfulness, justice, courage) practices would not be able to withstand the corrupting power that institutions exert. This is problematic not only because the aims of practices are not achieved. There is much more at stake, because practices are the vehicles through which the common good and the potential of human beings are actualized.

Social practices are the building blocks of larger social structures. For example, a university education involves not only practices of research and lecturing, but also commencement ceremonies, sporting events, accreditation, etc. Many of its practices are defined by a set of rules that are prior to the behavior of the participants: a PhD student may receive a hood from a professor, but it only counts as "hooding" within the set of rules that constitutes a hooding ceremony. At the same time, the practice offers participants roles to occupy, norms to follow, and reasons for actions: the professor has a reason to wear academic regalia, because it is required when participating in the ceremony. Complex, rule-governed practices depend on coordinated intentions and behavior (e.g., ceremony), involve accountability, and explicitly include judgments of correctness and incorrectness, while simple practices consist of patterns in behavior that result from social learning and cultural schemas internalized through socialization (e.g., the exchange of gestures). These can be prelinguistic bases for rule-following, with implicit, vague, and evolving norms such that behavior in accordance with them only requires basic responsiveness, not full-blown reflective judgments.

A criticism of a social practice can take two forms, depending on the norms it appeals to. In the case of *external criticism*, the standards employed stem from outside the practice criticized and it is always a possibility that the participants of the practice may not accept them. As Karl Popper (2000, 29) puts it, external criticism "attacks a theory from without, proceeding from assumptions or presuppositions which are foreign to the theory criticized."[18] For example, when critics appeal to

[18] Popper (2000) mainly discusses forms of criticism with respect to theories, but his considerations on a more general level apply to practices too.

human rights or the Bible, as some do in their criticism of medical practices, they are engaged in a form of external criticism. Here, it is irrelevant whether or not the criticized practice shares these standards, and if participants in the relevant practice do not accept those external norms, or do not think they apply, then they will probably not be very impressed by the criticism.

By contrast, *internal criticism* proceeds from the inside, employing standards that are seen as internal to the practice criticized, even if these are not explicitly recognized by all participants. The reference points are norms of the practice, not sets of beliefs shared by the participants of the social practice. Because it appeals to norms that the practice is seen as committed to, internal criticism is often seen as an effective form of criticism: judging a practice against its own standards does not face the difficulty of having to demonstrate the legitimacy of applying an external standard. As the standards appealed to are internal to the practice, raising awareness of their violation will likely be accompanied by some degree of motivation to change.

Popper (2000, 29–30) expresses reservations that such criticism "is relatively unimportant" since it must limit itself to pointing out inconsistencies within a practice. However, because theories as well as practices are also attempts at solving a problem, they can be submitted to internal criticism, for example, for failing to offer a solution or for failing to offer one that is superior to its competitors. In this way, immanent criticism may point out serious weaknesses even if the practice is internally consistent. Internal criticism is thus not necessarily conservative, solely aiming to restore or create internal consistency between norms and aims. In some cases, the fact that some norms of the practice are not satisfied stems from the fact that they are contradictory in themselves: they cannot be or are unlikely to be fulfilled for structural reasons (Jaeggi 2018).[19] Such a contradiction can arise if a practice embodies mutually opposing aims and norms that cannot be realized without contradiction or turn against the original intentions of the practice if realized.

In addition, we may distinguish two types of internal criticism. Internal criticism may target a norm that is *applicable* to a practice, or one that is *constitutive* of it. There are of course a large number of norms internal to practices, but some of them are somehow "privileged," picked out as the ones that ought to be conformed to (Brandom 1994, 28). Some of these

[19] This is often referred to as "immanent criticism" in the literature, particularly in the tradition of critical theory. The chapter will not make this additional distinction for purposes of simplicity.

norms governing the activity are constitutive norms, in the sense that the practice would not be the same without them, and specifying actions and roles that could not exist outside of the activity in question (nurse, doctor, etc.). The internal norms of practices need not be explicit but are often a mixture of more and less conscious and explicit elements (see Brandom 1994, ch. 1).

1.5.2 The Internal Criticism of Medicine

Let us now consider the criticism of medicine described in this chapter in light of this brief sketch of different forms of criticism. First, what unites these forms of criticism is their *internal* character. They all implicitly assume that medicine's own norms and values fail to be fully realized in the current institutional settings. Instead of condemning medicine by deploying independently justified standards (e.g., faulting medicine for rising expenditures or for failing to contribute to social justice) the criticism maintains that medicine has diverted from its course; it is no longer on the path toward its aim, and is thus failing to represent the values and norms it embodies as its own.

Second, medicine is criticized as *a social practice* that comprises both medical science and clinical practice. It coordinates a community in producing and using knowledge for the benefit of health, assigning roles for a large variety of participants (patients, nurses, physicians, etc.) in a variety of settings (e.g., the lab, the hospital, the clinic), all of which is governed by norms and social meanings internalized through participation. The criticism in particular appeals to two types of internal norms. The skeptical criticism mainly refers to the violation of *epistemic norms* of systematic knowledge-seeking (such as failing to communicate negative results) that is internal to medicine qua being science. The criticism of objectification, motivated by subjective experiences in health care settings, claims that objectification violates internal *moral norms* in medicine that govern the care of patients. Finally, the criticism of overmedicalization appeals to mixed sources. In some cases, the criticism is external, maintaining that overmedicalization is reproachable because it masks the social sources of suffering or because it contributes to the increase in the number of healthy people who are seriously concerned about their health. But, in most cases, the criticism is internal: the use of medical resources to address social or existential problems is not consistent with internal norms of medicine.

Third, the criticism appeals not merely to norms that are applicable to the practice, but to *constitutive norms*, understood in the sense that their violation is taken to undermine something that defines the practice. When O'Mahony laments the "corruption of medicine," it is conveyed that something definitive in medical research and practice has been lost. In most cases, the corruption of medical research, being implicated in overmedicalization, and the lack of compassionate care driven by patient need are taken to violate norms that not merely happen to be associated with medicine, but without which medicine turns into something else. At the same time, the criticism conveys that the *aim* associated with this practice cannot be achieved without adhering to these constitutive epistemic and moral norms.

1.6 The Use of the Criticism

Useful criticism tends to illuminate its subject, and metacriticism that systematically considers different strands of criticism can offer further contributions in this regard. We have so far been able to show that we are predominantly dealing with instances of internal criticism that appeal to constitutive norms of medicine, many of which are implicit. In general, implicit norms can be hard to identify, as we often first become conscious of their existence when they are violated. By conveying the perception of norm violations, the criticism makes important steps toward making more or less implicit norms explicit, which enables subjecting them to rational scrutiny. Moreover, we have also seen that the criticism appeals to two kind of norms. The skeptic's criticism appeals to *epistemic norms* of science, the criticism of overmedicalization appeals to norms governing medical knowledge that forbid certain uses, and the criticism of objectification appeals to *moral norms* that forbid a certain way of treating patients, even if their diseases are successfully removed. In the latter case, the criticism is informative in an additional way, because it shows that norm violation gives rise to "reactive attitudes" (e.g., indignation). Such reactive attitudes are best explained by positing the presence of implicit moral norms that are perceived to be violated.

These findings offer us a better view of the normative sources of the criticism, but they also help us comprehend that the criticism converges on more fundamental questions about (a) the aim of medicine, (b) the nature of medicine, and (c) the key concepts of health and disease, which correspond to the three levels of analysis of the normative approach that will be introduced in Chapter 2. With respect to (a), when critics like O'Mahony call on medicine to change its course and charge that it has overextended its dominion, this is based on a persuasion that medicine is currently not advancing toward its true aim. The criticisms of overmedicalization and objectification both point in this direction, maintaining that medicine has deviated from its course.[20] Moreover, the implicit assumptions of different strands are conflicting: the charge of overmedicalization seems to assume that (i) the aim of medicine is the removal and prevention of disease, while the charge of objectification seems to assume that (ii) the aim of medicine is to enhance well-being in a wider sense. If it turns out that (i) is true, then much of the charge of objectification looks unreasonable. After all, the successful removal and prevention of disease does not necessitate eliminating the features that critics of objectification draw attention to. In contrast, if (ii) is true, then the charge of overmedicalization begins to look mysterious.

With respect to (b), the skeptical criticism of medicine as science claims that current research practices are not consistent with the (scientific) nature of medicine. But Stegenga's skeptical thesis also has implications for questions about (a). This is due to the fact that although some of the arguments could be extended to domains of medicine, his thesis focuses on one kind of therapeutic intervention, namely intervention using pharmaceuticals. It does not systematically consider other types of standard interventions (e.g., surgical interventions, interventions in the form of radiation therapy or physical therapies, nonpharmaceutical rehabilitation procedures, lifestyle interventions). Moreover, in order for a medical intervention to qualify as effective, Stegenga's framework requires that it must target the constitutive causal basis of a disease, the harms caused by the disease, or both (Stegenga 2018, 15). This means that interventions that target conditions that are not "genuine diseases" (e.g., interventions on predisease states or on inappropriately medicalized conditions) are excluded. In addition, interventions in the form of vaccination are excluded because they aim to prevent the transmission of diseases rather than treat diseases (2018, 179), while a large number of other interventions (e.g., contraception, abortion, relieving teething pain, or menstrual cramps) are excluded because they do not target the constitutive causal basis of a disease or the harms caused by it.

Anticipating the objection that his view builds on an overly narrow account of the goal of medicine, Stegenga (2018, 52–3) grants "the multifaceted goals of medicine and the plural activities of physicians," but stresses that his analysis applies to *one* goal in medicine, which is the improvement of health by intervening on disease. While this seems like a suitable reply to the objection, we may note that the consequences of the

[20] Of course, this does not imply that medicine needs to return to some earlier era in which it succeeded in realizing this aim.

skeptical thesis for an overall assessment of medicine will depend on what the overall or final aim of medicine is and on how the goal to which Stegenga's analysis applies is related to it. For instance, the consequences of accepting the skeptical conclusion with respect to an overall assessment of medicine will be very different if one sides with critics of overmedicalization (i.e., the aim of medicine is the removal and prevention of disease) or if one sides with critics of objectification (i.e., the aim of medicine is to enhance well-being in a wider sense). In fact, critics of objectification could accept the skeptical conclusion while still holding on to the view that medicine as a whole is successful and produces significant progress.

Finally, with respect to (c), the charges of overmedicalization and objectification also implicate the notions of health and disease and claim that these have been altered to fit objectives that do not align with the aim of medicine. To assess whether such a claim is justified will likely require careful explication of the concepts of health and disease in light of the question about the aim of medicine.

Overall, there is therefore support for thinking that the criticism converges on such fundamental questions that correspond to the three levels of analysis of the normative approach. However, while these assumptions offer a normative backdrop for much of the criticism, critics do not offer systematic defenses of them. But, without this, the scope and significance of the criticism are limited. To better comprehend the criticism and to assess whether it is justified, the additional step of clarifying these fundamental issues seems indispensable. At the same time, attaining clarity about these issues makes a further inquiry worthwhile for additional reasons.

First, criticism that illuminates its subject can offer clues to the solution of the problem that it points to. However, taking further steps toward a solution would be greatly facilitated by an accurate account of the nature of the problem, which requires discerning the aim of the practice criticized. Without it, it is not clear what kinds of solutions are suitable with respect to the norm violations that propel the criticism. For example, by making a connection between aim and norms, one can discern whether the purported norm violation is an expression of a *local problem* (e.g., the norms of a practice no longer promote its aim) or a *systemic problem* (e.g., the norms of the practice are inconsistent). In the former case, problems typically have internal solutions, while in the latter they might resist a resolution within the current constellation.

Second, the emergence of the criticism is in part an expression of a new uncertainty about the proper role and scope of medicine in modern societies. Therefore, answering questions about the aim of medicine will not only help address the challenges that the criticism raises (i.e., determining the scientific nature of medicine, its proper boundaries, and the appropriate use of medical means), but it will also provide impulses to redefining medicine's role in society in the twentieth century. In this regard, consistent with the normative approach, it is important to stress that seeking to discover what the aim of medicine *is* should not be viewed as a separate undertaking from seeking to answer the question of what it *ought to be*.

Having offered reasons for why completing this additional step would be valuable, we may close by adding that the criticism also points toward how the question about the aim of medicine is to be answered. Taking seriously all three strands of criticism, one could argue that even if medicine suddenly became much more successful in curing and treating diseases (thus defeating the skeptical criticism), that would not resolve the criticisms of overmedicalization and objectification. And if the latter criticisms implicate questions about the aim of medicine, then we might start suspecting that there is more to medicine's aim than curing and preventing diseases. We will return to this issue in Chapter 6.

1.7 Conclusion

This chapter directed its focus at dominant forms of criticism, attempting to offer a better comprehension of their normative character and the challenges they convey. It was argued that the criticism is *comprehensive* (i.e., raises questions about both medical science and medical practice), mainly *internal* (i.e., relies on standards of evaluation that are assumed to be internal to medicine), and converges on a larger question about (a) the aim of medicine, (b) the nature of medicine, and (c) key concepts in medicine.

The criticism unearths challenges to medicine that require us to address basic questions on all of these three levels. Directing attention to these basic issues will not only help clarify to what extent the criticism is justified, but also assist an informed deliberation about the future of medicine. The main aim of the following chapters is to undertake this task. While Chapters 4 and 6 are chiefly dedicated to (a) and (c), Chapter 3 starts by addressing (b), and thus the question about the (scientific) nature of medicine. But, before that, Chapter 2 presents a particular, *normative* approach to philosophy of medicine that guides the inquiry in this book. It will be shown how this approach and the *three levels of analysis* it emphasizes can contribute to addressing the current challenges.

CHAPTER 2

Toward a Normative Philosophy of Medicine

2.1 Introduction: Philosophy of Medicine

What is the appropriate theoretical framework to address the basic questions that the different forms of criticism raise? The main aim of this chapter is to outline a particular *normative* approach to *philosophy of medicine* that can assist an informed deliberation about the future of medicine. The approach will distinguish between three levels of analysis that correspond to the three main questions that Chapter 1 identified and that current challenges to medicine raise. In a nutshell, the approach deserves the label "normative" because it is focused on uncovering implicit norms in medicine (e.g., with respect to the aim and nature of medicine), reflecting on to what extent they are actually fulfilled, offering corrections based on these findings, and evaluating the merits of current criticisms of medicine. But, first, what is philosophy of medicine?

Much philosophical work is dedicated to investigating foundational problems in particular sciences and their implications for broader philosophical questions. Major scientific fields are accompanied by subdisciplines of philosophy that examine the characteristics, aims, and methodologies of that field's subject matter. The "philosophy of medicine" is a subdiscipline of this kind, dedicated to investigating metaphysical, ethical, and epistemological issues in medicine (Marcum 2008, 8). While many salient philosophical questions with respect to medicine are located in the realm of ethics, where progress continuously raises new dilemmas,[1] philosophy of medicine is often seen as limited to epistemological, metaphysical, and methodological aspects of medicine (Caplan 1992, 69), which is why it is perceived as a different enterprise than medical ethics or health policy.

[1] Pressing questions of this nature include: Is it morally acceptable to genetically enhance or engineer human embryos? Or to ration expensive medication? Or to use the market to allocate organs for transplantation?

The growing interest in philosophy of medicine is not entirely surprising given that medicine has become pervasive in many aspects of our lives and given the numerous ethical issues raised by progress in medicine.[2] But, just twenty-five years ago, some doubted that it qualified as a proper field of study, because it (1) was not integrated into a cognate area of inquiry (e.g., like biochemistry with either biology or chemistry), (2) lacked a canon (i.e., a set of essential books, articles, and case studies), and (3) lacked certain puzzles, problems, and challenges to define its boundaries (Caplan 1992, 72–3). Today, philosophy of medicine meets these requirements. First, the philosophy of medicine is now integrated, and some argue that central problems in philosophy of medicine correspond to those debated in related areas in the philosophy of science.[3] Second, a brief look at the vast amount of material published in recent years reveals that philosophy of medicine now has the type of canon that characterizes distinct fields of investigation.[4] Third, as to boundary-defining problems, puzzles, and challenges, it suffices to highlight debates about the nature of functions, the meaning of the concepts "disease" and "health," or biological versus mental levels of description in psychiatry.

Owed, in part, to this dynamic development, discussions about the objectives and methods of philosophy of medicine continue. By presenting and defending a particular *normative* philosophical engagement with medicine that guides this book, this chapter also contributes to the general debate, in part by rethinking some of the basis of philosophical engagement with medicine – a task that requires addressing fundamental questions about the nature of philosophical inquiry itself and its relation to science.

[2] The reasons for such growth are typically diverse and multifaceted, often having more to do with factors external to the area (the social dynamics of research, the politics of research funding, media coverage, etc.) and not always related to developments internal to the area, like some sort of progress or breakthrough.

[3] For example, findings in evidence-based medicine (EBM) and epidemiology use large samples of population data, which raises questions about the kind of evidence that statistical methods can offer. Some argue that they can offer causal knowledge, while others argue that, lacking experiment-based evidence of a potential mechanism, they cannot provide us with causal knowledge of either a disease or the efficiency of a drug (Russo and Williamson 2007; for a survey, see Huneman, Lambert, and Silberstein 2015; Daly 2017).

[4] The last decade witnessed the publication of four introductions to the philosophy of medicine. During the same period, four major handbooks were published: the *Springer Handbook of the Philosophy of Medicine* (ed. Thomas Schramme and Steven Edwards 2017), *The Routledge Companion to Philosophy of Medicine* (ed. Miriam Solomon, Jeremy Simon, and Harold Kincaid 2017), *The Bloomsbury Companion to Contemporary Philosophy of Medicine* (ed. James Marcum 2016), and the *Handbook of Analytic Philosophy of Medicine* (ed. Kazem Sadegh-Zadeh 2012) (see also Daly 2017). Meetings of the Philosophy of Science Association now dedicate plenary sessions to philosophy of medicine (Lemoine, Darrason, and Richard 2014).

2.2 Normative Philosophy of Science

In order to outline a specific philosophical engagement with science that focuses on normative aspects, we may accept some version of the standard view of philosophy of science as a second-order discipline that deals with sciences that (at least in paradigmatic cases) study the natural world. Accordingly, in both philosophy and science, theories are developed to secure evidential support with data, but the empirical data in the sciences are about the natural world (at least in paradigmatic cases), whereas empirical data in philosophy are about the natural sciences. But, while accepting this general view, we can simultaneously hold onto a view stressing that second-order philosophical inquiry is *continuous* with what is already more or less explicitly present in the sciences. In this sense, as Elliott Sober (2008, xv) puts it, "normative philosophy of science is continuous with the normative discourse that is ongoing within science itself."

Like Sober's own approach, the kind of normative philosophy of science gestured at here engages in an evaluative endeavor with the aim "to distinguish good science from bad, better scientific practices from worse" (Sober 2008, xiv). Of course, such an approach is not entirely new. Philosophers have done normative work, making recommendations about how science ought to proceed and how its concepts ought to be understood. Increasingly, such work proceeds based on extensive descriptive analyses of scientific practices and not merely on abstract theories from philosophy (Kaiser 2019). But, while normatively contentful work in philosophy of science (e.g., in Sober's work) made explicit standards by which scientific theories ought to be evaluated, it has focused primarily on issues in epistemology and metaphysics. Only relatively few philosophers like Helen Longino and Philip Kitcher have continuously engaged in normative work that also takes into account the normative layers in science that stem from its social context and its embeddedness in societies.

Building on their work, a fully developed normative philosophy of science would pursue the task to offer a comprehensive account of normativity as an integral part of science. However, given the aims of this book, the normative approach suggested here has a narrower focus on three levels of normativity that intertwine in particular scientific fields. Corresponding to the types of questions that current challenges to medicine raise, we distinguish between three levels of analysis: the aim of science, the nature of science, and key concepts.

2.2.1 The Aim of Science

Questions at this level concern, for example, the aim of science in general and the more particular aim that propels specific scientific fields. The articulation and critical assessment of such aims are continuous with scientific activity, but benefit from philosophical expertise, which may for instance help bring implicit assumptions and norms to the surface. Identification of these aims is decisive for an informed reflection about the direction in which science should proceed. First, setting the general priorities of scientific research turns on questions about what a society most needs to know, and that requires taking a stance on the most fundamental needs, aspirations, and values of the societies in which the research is embedded. Philosophy can make a crucial contribution here, assisting in debates about the appropriate aims of science. Second, the epistemic aim of science (whether it is truth, knowledge, understanding, or something else) will set different standards for what counts as progress. For example, if science follows the fundamental aim to offer truth about the world, then its progress is assessed in terms of how successfully this aim is realized. But, given the vast number of truths attainable, it is obvious that science selects what are the *important* truths worth pursuing. Because this process involves reflecting on reasons for what it is that makes those truths valuable, philosophical reflection on the relevant epistemic values (e.g., truth-conduciveness, avoiding falsity) and practical values (e.g., agency, autonomy) will be decisive for comprehending progress in science.

2.2.2 The Nature of Science

At this level, philosophical inquiries can, for example, concern the nature of scientific knowledge and the methods of acquiring knowledge in science. Science is seen as offering a reliable basis for decision-making not because of the epistemic and moral virtues of scientists, but because science pursues a knowledge-producing inquiry that is superior to commonsense methods. A step toward determining how best to achieve the aim of scientific inquiry is attaining more clarity on what it is that demarcates science not only from commonsense knowledge, but also from pseudoscience. To effectively pursue the aim of science, concerns that can be located on this level include addressing the nature of the epistemic norms of scientific inquiry. These distinctions can productively interact with answers to questions about the aim of science. For example, if we characterize the nature of science by certain epistemic norms, then this also raises the question to what extent these align with practical norms that derive from the aim of scientific inquiry.

2.2.3 Key Concepts

At this level, the task is not merely to describe the use of the relevant concepts in scientific practice, but to offer critical reconstructions and perhaps revision. While one can find it sufficient to offer a descriptive analysis of, for example, causal and explanatory claims in the sciences, perhaps clarifying their criteria of application, one can also use descriptive analysis to make normative "recommendations about what one ought to mean by various causal and explanatory claims" (Woodward 2003, 7). Moreover, descriptive analysis may reveal that certain concepts used in the sciences are not as neutral with respect to values as it may first seem. For example, the term "validity" might be descriptive (i.e., picking out features of the world), evaluative (i.e., providing reasons for action), or both, expressing a "thick concept" with both evaluative and nonevaluative content. On the last possibility, "validity" might be akin to "murder," thus expressing a concept that has evaluative content (e.g., wrong) but also nonevaluative content (e.g., premeditated killing). The concepts "health" and "disease" could function in the same manner. For example, "disease" might express a concept with both evaluative content (e.g., harm) and nonevaluative content (e.g., part-dysfunction). Exposing these kinds of normative features can be a valuable contribution to scientific inquiry.

The approach is not only normative, but also *revisionist*: it not only aims to reflect on how we ought to reason in light of extant conceptual structures, but also considers how those conceptual structures *ought* to be. Deviating from standard conceptual analysis, Rudolf Carnap introduced the term "philosophical explication," the task of which "consists in transforming a given more or less inexact concept into an exact one or, rather, in replacing the first by the second" (Carnap 1950, 3). The aim of explication is thus not to discover meaning, extension, or criteria of application while respecting conventional usage, but providing alternatives that are superior to the existing concept (Loomis and Juhl 2006; Kitcher 2008; Schupbach 2015). Carnap's account has served as a model for contemporary work under the label of "conceptual engineering," which goes beyond analyzing extant concepts to assessing whether these concepts are the best tools for understanding the relevant aspects of reality and considering revision or replacement in case the answer is in the negative (Haslanger 2000; Burgess and Plunkett 2013; Eklund 2014; 2015;

Cappelen 2018). To be clear, the claim is not that revisionism is unique to philosophy. Scientists often replace concepts (e.g., concepts of folk biology) with concepts that better suit their theoretical purposes (Eklund 2014). Nonetheless, qua conceptual work, philosophy has a particular expertise in this area (Cappelen 2018, 5; Chalmers 2020).[5]

Stretching Carnap's framework, we may distinguish between *epistemic* and *emancipatory revisionism. Epistemic revisionism* offers justification on epistemic grounds. It is thus propelled by an effort to improve concepts that suffer from the deficiency of "inexactness" and therefore lead to less true and illuminating generalizations.[6] Using an example from zoology, Carnap (1950, 5–6) highlights the replacement of the vague and broad concept "fish" by the scientific concept "piscis," which is more circumscribed and refers only to cold-blooded aquatic animals that have gills throughout their life. "Piscis" have more properties in common than "fishes," and there are more illuminating and true generalizations that involve "piscis" than the explicandum "fish."[6] The explicatum "piscis" is therefore productive and represents an epistemic gain.[7] To take an example from philosophy, some have argued that the notion of "free will" is incoherent (van Inwagen 2008, 327–8) and could be taken to express closely related, but different, properties. In cases of "conceptual pluralism," contradicting claims are not necessarily a disagreement, but a verbal dispute (Chalmers 2011).

Emancipatory revisionism offers justification for revision on nonepistemic grounds. For example, some terms appear to serve the goals of explanation and prediction (e.g., natural kind terms), but they actually furnish

[5] Chalmers (2020) thinks that conceptual engineering is broad and covers *de novo conceptual engineering*, which does not particularly try to fix or replace other concepts. Examples include consciousness access, epistemic injustice, and belief. He distinguishes homonymous (same-word) and heteronymous (different-word) conceptual engineering, and endorses a form of conceptual pluralism (different things that deserve to be called "genes"), holding that there are many interesting concepts in the vicinity of philosophical terms and that the task of conceptual engineering is to articulate those roles and to identify the concepts that are best suited to play these roles.

[6] A similar story can be told about the replacement of "salt" with "NaCl," and "warm" and "cold" by "temperature." Carnap (1950, 7–8) puts forward four conditions for a successful explication and these examples meet all of them. According to the conditions, the explicatum should (1) retain similarity to the explicandum, (2) have an exact specification within a system of scientific concepts, (3) be epistemically productive (i.e., allow articulating numerous universal statements), and (4) be as simple as possible.

[7] For example, some recent work maintains that paradoxes and infertile debates are sometimes symptoms of an underlying defect in the concept itself. Some argue that "truth" gives rise to paradoxes and contradiction not because of faulty reasoning or premises, but because "truth" is incoherent. For this reason, it should be replaced by "ascending truth" and "descending truth" (Scharp 2013).

pseudoscientific legitimacy to social practices that sustain domination and oppression. Consider, for instance, "race." Some have argued that it presupposes "racialism," according to which human kind naturally divides into racial groups, the members of which inherit cognitive, emotional, physical, moral, and cultural features that they do not share with members of any other race (Appiah 1998). On such a basis, *epistemic revisionism* would conclude that since racialism is empirically false, the term "race" has an empty extension and might be eliminated. Alternatively, one might argue that existence of races not only presupposes "racialism" but also "racism" (Van den Berghe 1967). The objective physical differences tied to skin pigmentation are devoid of social significance, and they only become relevant if something like racism marks and elevates these differences. In such a case, conceptual analysis can expose that "race" has functions that conflict with values that we endorse, making it a target for *emancipatory revisionism*. The revision in such a case is not primarily legitimated by epistemic deficits, but by considerations about what the concept ought to be or what extensions it ought to have, for example, in order to promote social justice.

For an example from the history of medicine, consider "madness." It may seem to be a natural kind term that primarily functioned to help diagnose mental disorders. This would presuppose a view according to which humankind naturally divides into groups, and if this is empirically false, then epistemic revisionism might conclude that the term is best eliminated. Alternatively, we may find that the concept was motivated by repressive regimes and assisted in legitimizing and lending medical authority to practices of social exclusion. In that case, emancipatory revisionism might enter the picture.[8]

Overall, the normative approach to philosophy of science outlined here explores science not merely as what it is and how it proceeds, but also as what it ought to be and how it ought to proceed. It is not merely interested

[8] There are many other examples. For example, "terrorism" belongs to a classification scheme that groups together categories of violence that share common characteristics, so it appears that it helps achieve the cognitive task of reducing complexity by drawing the boundaries of a certain type of violent instrumental action, much like "burglary" and "armed robbery." But, this is not what "terrorism" is actually doing. The fact that "terrorism" is not applied to all cases of violent crime with political or social objectives (including white violence) not only makes it less useful, but also indicates that it has no neutral application. In such cases, justification for revision is ethical or political, based on the idea that the new usage would be good for various ethical or political reasons. Or, consider "woman," which seems to express the concept of an adult human female. Sally Haslanger (2000) has argued that "woman" is primarily used to subordinate people based on their stereotypical female characteristics. Haslanger's project aims to oppose subordination by making the concept explicit.

in normativity, but in a comprehensive approach to normative matters that includes several levels. To avoid misunderstandings, it is crucial to emphasize two features. First, focusing on these levels of normativity is not adopting a privileged viewpoint outside of science. Rather, these reflections are *continuous* with what is already implicit in the practice of science. Second, the normative approach need not exclude traditional themes in philosophy of science from its domain of interest; rather, it aims to understand them within a larger framework of science in society and questions about values that are central to philosophy of science.

2.3 Normative Philosophy of Medicine

Such a general approach is particularly relevant to philosophy of medicine and contributes to ongoing discussions about the objectives and methods of philosophy of medicine. It yields an approach to philosophy of medicine that has a broader scope than some of its founders envisaged. Deploying such a normative approach and its levels of inquiry to medicine, we arrive at a range of key areas and questions that have a different profile than traditional approaches to philosophy of medicine, but that are well suited to help reach our aims of this book.

(1) *The aim of medicine.* What is the aim of medicine, and what are the relevant criteria for progress?

(2) *The nature of medicine.* To what extent can medicine be said to be a science that displays similarities with respect to the type of knowledge it produces and the methods it deploys for this purpose?

(3) *Key concepts in medicine.* What is the status of key concepts like "health" and "disease"? Are they suitable for the purposes of scientific inquiry in their current form, or is there is reason for (epistemic or emancipatory) revision?

Overall, this book not only applies such a normative approach to the philosophy of medicine, but is in many ways organized around these three levels and areas, as they are deemed of key importance for addressing the contemporary challenges to medicine mentioned in Chapter 1. First, strengthening the case for the suitability of this approach for addressing the main questions and challenges identified in Chapter 1, it is not difficult to see how these three levels of analysis map onto the conclusion in Chapter 1, which maintained that dominant forms of criticism converge on a larger question about (a) the aim of medicine, (b) the nature of medicine, and (c) key concepts in medicine.

Second, very much in line with Sober's suggestions, the normative approach offered here is continuous with the internal normative discourse in medicine that Chapter 1 scrutinized. As was shown, the different forms of criticism are united by their *internal* character: instead of condemning medicine by deploying independently justified standards, the criticism maintains that medicine is no longer on the path toward its aim, fails to represent the values and norms it embodies as its own, and likely operates with overly permissive notions of health and disease. Thus, while the criticism relies on standards of evaluation that are internal to medicine, the normative approach suggested here further elaborates on them by using the resources of philosophical work.

In closing, we briefly compare the normative approach to established approaches in philosophy of medicine. First, the normative approach resists the traditional model, according to which "philosophy of medicine is a separate and separable entity from philosophy of science" (Pellegrino 1998, 326). Moreover, the traditional model also claims that philosophical work on medicine is "detached from the method and content of medicine" (Pellegrino 1986, 13) and "ought to examine its subject, medicine, from a perspective and methodology not contained within that subject, itself, or its component disciplines" (Pellegrino 1986, 10; 1998, 325–6). The underlying idea holds that philosophy is discontinuous with science, and along with this division, Pellegrino and followers have distinguished between *philosophy of medicine* and *philosophy in medicine.* While the former deals with broader questions about the aim of and values in medical activities, the latter proceeds by applying the tools of philosophy to medically defined problems like causality, the logic of diagnosis and prognosis, and the metaphysical presuppositions of concepts of disease and health (Pellegrino 1976; 1986; 1998). Thus, using our distinction between three levels, Pellegrino largely restricts philosophy of medicine to the first level, while assigning issues on the second and third levels to philosophy in medicine, which he thinks of in terms of traditional philosophy of science. Moreover, he also restricts the scope of philosophy of medicine to clinical practice, due to a narrow view of what science is. While the distinction (i.e., philosophy of medicine versus philosophy in medicine) relies on the narrow notion of science that Chapters 3 and 4 take issue with, the view proposed in this chapter encompasses issues and questions that others have assigned to philosophy *in* medicine. As we have seen, it is somewhat artificial to entirely separate first-level questions (e.g., about the aim of medicine and the notion of progress that applies to it), from second-level issues (e.g., about the methods that medicine applies to reach its aim), and third-level matters (e.g., the nature of health and disease).

Second, the normative approach also resists a more recent model, which comprehends philosophy of medicine as "the study of epistemological, metaphysical and methodological dimensions of medicine" (Caplan 1992, 69). This view differentiates philosophy of medicine from normative enterprises such as medical ethics or health policy and aligns philosophy of medicine with philosophy of science. This is reflected in an introduction to the topic by Thomas Schramme, maintaining that philosophy of medicine regards "medicine from a theoretical point of view, i.e., in order to analyze, understand, or explain aspects of the theory and practice of medicine. Bioethics, in contrast, discusses normative problems in medicine from a practical point of view, i.e., in order to provide guidance as to how people should act" (Schramme 2017b, 4). This distinction between a theoretical and a practical perspective, one aiming at analysis and the other at normative guidance, does not imply, as Schramme clarifies, denying issues regarding value and morality in philosophy of medicine. Nonetheless, this is precisely what characterizes the dominant orientation in the literature, which has been guided by a traditional understanding of the tasks of philosophy of science (i.e., focusing on epistemology and metaphysics). Due to such a focus, apart from a limited number of exceptions (e.g., Solomon 2015; Stegenga 2018; Broadbent 2019), normative issues that stem from medicine's embeddedness in societies were predominantly assigned outside of the purview of philosophy of medicine.

It is here that the normative approach widens the scope of philosophy of medicine. Enabling a critical reflection on medicine as a whole, on research and practice, normative philosophy of medicine not only tries to identify aims, values, and norms in medicine, but also brings them under critical scrutiny and normative assessment. It holds that ethical and policy issues in medicine cannot be comprehended fully without considering the nature and aims of medicine, which, in turn, cannot be pursued in isolation from questions about medicine as a social institution embedded in today's societies.

2.4 Three Objections to Continuity

The normative approach displays important advantages compared to traditional accounts, and it is well positioned to help examine the aim and nature of medicine and to reach the goals of this book. However, it relies on comprehending the relationship between philosophy and science as continuous forms of inquiry. Call this the Continuity View. It stresses that the normative approach on offer is continuous with the internal normative discourse in medicine, but it also denies that the methods used in philosophy are unique to philosophy and that they grant access to some distinct realm of truth that is inaccessible to science. Thus, it opposes the view adopted by Pellegrino, according to which philosophical work on medicine is discontinuous with medicine in terms of method and content and must study its subject with methods not contained within that subject (Pellegrino 1986; 1998; 2001). Instead, the Continuity View sees philosophical inquiry as deploying tools that it shares with scientists (e.g., rigorous argumentation, logic, conceptual analysis), but philosophers operate them with a special expertise and meticulousness (Laplane et al. 2019, 3950). Nonetheless, the Continuity View does not imply the stronger thesis that philosophy *is* science or that all philosophy is continuous with science.

The term "continuity" is often associated with Willard Van Orman Quine's work, which is motivated by his rejection of the analytic–synthetic distinction and by his denial of any a priori ground ("first philosophy") outside of science that may justify it. This chapter uses "continuity" in a less specific sense and does not consider these debates. Others might associate "continuity" with demanding versions of naturalism, but what is suggested here is only naturalist in broad terms: it highlights that philosophical inquiry is not restricted to meta-theoretical activity from the armchair (as opposed to empirical observation) and that philosophy being continuous with science does not mean that it is entirely reducible to it.[9]

[9] For a discussion of various versions of naturalism, see Morganti (2016). In the philosophy of medicine, naturalism is most often associated with the work of Christopher Boorse on the concept of disease. More recently, Mäel Lemoine (2015) has criticized a tendency in philosophy of medicine to study the concept of disease based on "disease judgments" (rather than explanations) and a pre-naturalized notion of disease. Lemoine proposes "to distinguish a naturalist notion of disease from a naturalized notion of disease" (Lemoine 2015, 21) and suggests a shift from focusing on a common understanding of disease to investigating biomedical research for more general facts and theories that might help define disease. This suggests a different and more demanding kind of "continuity" than what is implied in the approach suggested here. That said, on a general level, Lemoine is right to harbor misgivings about the implicit science/philosophy division of domains. For Lemoine, this is a point about the definition of disease, but in the context of this book, a similar general point extends to questions about the aim and nature of science. These questions belong exclusively neither to philosophy nor to science. Others suggest a reorientation with respect to questions about "disease" by deploying the tools of experimental philosophy (De Block and Hens 2021; Veit 2021). Just like much contemporary work in philosophy, some of the aims in this book would benefit from evidence from experimental philosophy, but they do not require it.

But even so, proponents of the traditional model like Pellegrino, who see philosophical work as discontinuous with science, will likely object to the Continuity View. They could proceed by presenting three lines of argument. First, in order to support the thesis that philosophy and science are discontinuous, they could argue that science engages in *empirical* work, while philosophy engages in *conceptual* work, which does not aim at factual knowledge or empirical explanations. Second, they could argue that the claim that philosophy is continuous with science amounts to a problematic species of *scientism*. Third, to back the claim that philosophy and science are discontinuous, they could argue that scientific work is *descriptive*, while philosophical work is *normative*, essentially in the business of providing standards by which to evaluate its subject matter. Let us test these objections one by one.

2.4.1 *Empirical versus Conceptual*

The gradual separation of the sciences and philosophy is often described as giving rise to a *difference in type* between philosophical and scientific inquiries.[10] For example, discussing psychology breaking off from philosophy, Michael Dummett maintains that "what remains is a discipline that makes no observations, conducts no experiments, and needs no input from experience: an armchair subject, requiring only thought" (Dummett 2010, 4). Unlike the sciences, philosophy is not dedicated to describing the physical properties of entities, but, rather, the concepts we use to speak about them. While science investigates first-order aspects of nonlinguistic reality using observation and experiment, philosophy aims at establishing a priori necessary truths by deploying a characteristic method, conceptual analysis, which can be done from the armchair, independent of empirical assumptions, observations, experiments, or experience.[11]

In reply, we may start by noting that conceptual analysis is not unique to philosophy. It is deployed to define concepts such as prime numbers in mathematics or "placebo" and "endophenotype" in medicine (Kendler and Neal 2010).[12] Still, our opponents could emphasize that philosophy and

[10] For example, when biology developed into the science of living organisms, philosophical questions like what constitutes life (i.e., the question of whether life requires something in addition to those physical and chemical processes that are absolutely necessary for it) became perceived as a question that belonged to philosophy rather than to biology (Rosenberg 2005; Chalmers 2015, 21).

[11] The discontinuity due to a "difference in type," as A. J. Ayer (1946, 50–1) has put it, means that philosophical and scientific propositions "cannot conceivably contradict one another."

[12] Of course, concepts that constitute the targets of philosophical analyses are typically complex and deeply intertwined with and embedded in practices and institutional contexts. They are tied to

science are discontinuous because (a) being empirical is a necessary condition for science and (b) philosophy is not empirical. However, both (a) and (b) are flawed. As to (a), it builds on an erroneous view of science as exclusively interested in discovering empirical aspects of phenomena by using observational and experimental methods. There are a number of formal sciences (e.g., mathematical logic, geometry, arithmetic) that are recognized as proper sciences while not being empirical in the sense of being based on observation and experiment. This seems to indicate an important issue – unfolded in detail in Chapter 3 – namely, that being based on observation and experiment is not what makes a discipline science, but rather some kind of *systematic* and rigorous organization of the inquiry and the body of knowledge that it generates. Thus, if being empirical is not a necessary condition for science, then the claim that philosophy is not empirical is consistent with the claim that it is continuous with science.[13]

As to (b), even if we accept that being empirical is a necessary condition for science, there are at least two reasons for thinking that conceptual analysis might be properly experimental and thus qualify as empirical. First, consider that philosophers investigate concepts by examining their own intuitions about hypothetical cases. For example, a philosophical hypothesis can be tested by conducting thought experiments. Philosophers often imagine cases in which the conditions are fulfilled, attempt to produce counterexamples, and proceed by rejecting or amending the hypothesis. It may be argued that in this process, philosophers generate a body of empirical data ("intuitions") that they consult in order to test the hypothesis (McGinn 2015). In this sense, conceptual analysis may be experimental: the results of thought experiments can be

preconceptions in a very different way than the relatively pure and insulated concepts used in mathematics, which render it a truly aprioristic field of inquiry (Dennett 1986). But it is not clear why such differences would amount to methodological dissimilarities of a kind that would cause problems for the Continuity View. Opponents might add that the essential difference is that unlike mathematicians, philosophers analyze concepts as they are understood in everyday usage. This is indeed true in many cases, but this reply fails to take into account that there are plenty of examples in philosophy where conceptual analysis aims to capture concepts as used in professional contexts, outside of everyday discourse. For example, this is the path that Christopher Boorse chooses for his influential analysis of health and disease.

[13] Based on this line of thought, McGinn (2015) argues that philosophy is best seen as a formal science. But this view seems too strong. It requires accepting that philosophical work overwhelmingly consists of doing conceptual analysis, which is something that we have granted so far, but which actually does not seem to reflect the reality of the profession. But if philosophy is not overwhelmingly conceptual analysis, and sometimes deploys abductive inferences and inductive methods to identify regularities and laws from observations, then this undermines the idea that philosophy belongs to the formal and deductive sciences.

seen as new pieces of information that philosophers acquire by using their conceptual apparatus. The fact that the intuitions are generated by the experimenter *in* the experimenter may lead to objections that intuitions have the epistemic status of hunches and, hence, should not be taken as evidence, but this does not undermine their status as experimental. Similarly, some argue that armchair methods straightforwardly rely on empirical information (Kornblith 2017, 155), whereas others argue that they possess a subtle empirical dimension, but still offer low-grade a priori justification (Henderson and Horgan 2011, 2).

In reply, our opponent could accept these points, but still resist the Continuity View by stressing that philosophers aim at necessary truths, which can only be known a priori, whereas scientists aim at contingent truths that can only be known a posteriori.[14] But this is false and runs together necessary truth and a priori knowable truth. A metaphysically necessary truth could fail to be a priori, as, for example, in the case of the statement "Hesperus is Phosphorus," which is necessarily true, even though it can only be discovered a posteriori (see Kripke 1980, 21). Moreover, contingent statements can be a priori. Consider the following disjunction, which is itself a priori: *p* ("no one can be his or her own parent") or *q* ("nurses are stressed these days"). Mary knows *p* a priori (the concept of parent has everything that is required), but does not know *q* and lacks knowledge about the current conditions of nurses. Being a nurse himself, John knows a posteriori *q*, but not *p*. Both Mary and John are able to competently deduce the disjunctive conclusion. However, from the premise *p* Mary comes to know *q* a priori because the deduction is not dependent on experience. In addition, from the premise *q*, John comes to know *p*, a necessary truth, a posteriori.

2.4.2 Scientism

Proponents of the traditional model like Pellegrino might draw on an argument presented by Dummett, which argues that the claim that philosophy is continuous with science amounts to a species of *scientism*, or "the disposition to regard the natural sciences as the only true channel of knowledge" (Dummett 2010, 35). While this objection may proceed

[14] This draws on the well-known distinction between two specific ways of knowing. S knows *p* a priori if and only if S has come to know *p* independent of experience (except for the experience required to learn the meanings of *p*'s constituent term), and S knows *p* a posteriori if and only if S has come to know *p* in a way that depends on experience.

based on a narrow and restrictive or on a broad and permissive conception of science, neither of these possibilities turns out to be promising. Here is why.

If the charge of scientism uses a narrow conception of science, as constitutively based on observations and experiment, then it is inconsistent and clearly not what the Continuity View proposes. First, as we have seen, observation and experiment is not necessary for science. But even if it was, philosophy might qualify as empirical, at least if one accepts the previous portrayal of the nature of thought experiments. Second, on a narrow conception of science, it is not difficult to find questions that science is not equipped to answer, which means that science could not be the only channel of knowledge. For example, questions about the nature of causation, valid inference in medicine, and the ontological status of numbers do not admit answers found through observation and experiment. Third, scientism fails by its own standards, because it relies on a hypothesis about science that cannot be deduced by logical methods from elementary truths and for which there is no empirical (experimental or statistical) support. It suggests that we should not believe any proposition that cannot be scientifically supported, but then questions arise about why we should believe that very proposition itself, since it cannot itself be scientifically supported. It seems safe to conclude that if the objection is stated with a narrow conception of science in mind, then it fails.

If the charge of scientism uses a more permissive conception of science, according to which observation and experiment are not necessary conditions for science, then the reply follows a different path. On a more permissive conception, forms of philosophical work could in principle count as science. But, if so, the Continuity View could simply accept the label "scientism." This opens up another sort of objection, maintaining that philosophy is continuous with science, but that it (or at least a major part of it) is *bad science*. For such a claim to get off the ground, our opponent would have to make recourse to criteria for science other than observation and experiment. If so, then the opponent's best bet is probably going to be some epistemic virtue like systematicity. But then, as we shall see in Chapter 3, someone committed to the Continuity View need not disagree.

2.4.3 Descriptive versus Normative

Some opponents might highlight another option, according to which philosophy is a *normative discipline*, essentially in the business of providing standards by which to evaluate its subject matter (Thomasson 2015; 2017). Unlike the sciences, then, the primary aim in philosophical inquiry is not a descriptive one (e.g., *describing* how people actually reason about X, behave toward X, or evaluate X), but a prescriptive one (e.g., how they *ought to* reason, behave, and judge). Descriptive questions about what *is* the case are best dealt with via the empirical methods of the sciences, but these do not tell us what *ought to be* the case, which is what philosophy does.[15]

Indeed, normativity is involved in a large number of philosophical undertakings, not merely in certain parts of philosophy like aesthetics and normative ethics. Aristotle's doxastic law of noncontradiction does not describe how we reason, but offers norms for reasoning, which is why Frege urged us to understand logic similarly to ethics as a "normative science" (Steinberger 2016).[16] Parts of philosophy of science explore what methodologies should be employed to generate data and what conclusions should be drawn from them (Thomasson 2015). However, accepting the idea of philosophy as a normative discipline does not present a challenge to the Continuity View.

First, one may argue that it relies on a derived notion of normativity on which many sciences will qualify as normative. Consider Aristotle's doxastic law of noncontradiction, according to which it is impossible to hold the same thing to be *F* and not to be *F*. In itself, the law of noncontradiction is a descriptive claim from which it does not follow that one ought to think in conformity with it. One can derive explicitly prescriptive claims (you ought not believe that *x is F* and *x is not F) only by* adding a normative claim, namely that violating the law will prevent you from getting to the truth, which is a bad thing. Consequently, the normative force derives not from the logical principle, but from a statement from outside of logic. If normativity is understood in this derived sense, then normativity is

[15] Amie Thomasson's account is developed as a reply to the charge from Hawking and Mlodinow. Thomasson (2015; 2017) has argued that the criticism is correct if we think of philosophy as a factual discipline aimed at the truth about aspects of reality. She maintains that if philosophy is "on par" with the sciences, then we end up with a "proliferation of fanciful views that seem like either bad science or wild speculation, and no idea how to choose among them" (Thomasson 2015, 20). One the other hand, if one insists that philosophy's role lies in conceptual work describing how we think about various topics, we end up with a "a rather parochial and limited conception of philosophy" (Thomasson 2015, 20).

[16] Philosophical work on paradoxes and puzzles often points out premises that we ought to revise, while parts of epistemology are concerned with how we should reason under conditions of uncertainty, how we should acquire evidence, and when we should count someone as having knowledge.

ubiquitous and relatively uninteresting. For example, consider the claim that "medical scientists have established that antibiotics do not affect the course of the seasonal flu." Nothing normative follows from this claim, but we may – and typically do – add the normative claim from outside of medicine that one ought not take medication that is shown to be ineffective for a condition.

Instead of the derived notion, our opponents could make the stronger claim that philosophy is normative in a nonderived sense. The claim could take on different shapes, but one is that concepts in philosophy are normative in the sense that they can only be fully understood with recourse to normative terms. For example, we may understand validity in descriptive or normative terms. Understood in descriptive terms, we may say that (a) an inference from premises to conclusion is valid iff it is impossible for the premises to be true and the conclusion nevertheless to be false. Understood in normative terms, we may say that (b) an inference from premises to conclusion is valid iff one ought not to believe the premises without believing the conclusion. In contrast to (a), (b) implies that validity cannot be fully understood without grasping its normative implications, which directly reflects that accepting an argument as valid normatively constrains an agent's doxastic attitudes.[17] Without going into further questions about normativity in logic, the important upshot is that just like derived normativity, nonderived normativity is not restricted to philosophy.

As an example, consider the notion of evidence in discussions on evidence-based medicine (EBM), which can be analyzed in normative terms, and thus in terms of what conclusions we should draw from it. Thus, we may say that E is strong empirical evidence against a scientific hypothesis H iff one ought not to believe H while at the same time believing E. Accepting E as evidence normatively constrains an agent's doxastic attitudes, and grasping E's implications is necessary to understand the concept E as something that confers justification for belief. This brings to the fore that, at least with respect to justifying beliefs, "evidence" and "reason" (to believe) are quasi synonymous, the main difference being that the former cannot occur in the plural (a mass noun), while the latter can (a count noun) (Kelly 2016).

[17] This example draws on debates about whether logic can be a normative discipline (see MacFarlane J., 2004, unpublished manuscript https://johnmacfarlane.net/normativity_of_logic.pdf; Steinberger 2016).

2.5 Conclusion

To outline the type of philosophical engagement with medicine that guides this book, this chapter presented a *normative* approach to philosophy of medicine. It deserves the label "normative" because it uncovers norms linked to the aims, nature, and key concepts of medicine, assesses to what extent they are actually fulfilled in practice, and offers corrections based on these findings. Moreover, this approach also allows evaluating the merits of the current criticisms of medicine, and parts of the current chapter were dedicated to showing how its *three levels of analysis* can contribute to addressing the challenges to which Chapter 1 drew attention. The suggested approach is well suited to help reach the aims of this book, but through a critical dialogue with the views of Pellegrino and others, it can hopefully also make a contribution to the ongoing discussions about the objectives and methods of philosophy of medicine.

While the normative approach was developed in a critical dialogue with influential accounts within the philosophy of medicine, it was highlighted that it relies on the Continuity View, which acknowledges significant differences of degree between philosophy and science, but holds that they are continuous. The Continuity View is consistent with the ongoing and often fruitful interaction between the disciplines[18] and with the intuitively appealing view that in spite of numerous differences, both philosophy and science ultimately attempt to answer a more basic urge, namely, the desire to attain an increasingly profound and systematic understanding of the world (Hempel 1966, 2; Williamson 2018). Surely, there are clear methodological, stylistic, and other differences, and many scientific undertakings will not exhibit much resemblance to philosophical work. But these differences do not warrant denying continuity, especially since we find a similar diversity within those areas of inquiry that are traditionally acknowledged as proper sciences.

This chapter offered a defense of the Continuity View against objections that could be launched by proponents of a traditional view of philosophy

[18] For example, on the one hand, Fodor's work on modularity has been influential in cognitive science, while Dennett's work on our ability to think about others' false beliefs has been influential for development of the false-belief task, widely deployed in evaluating cognitive developmental stages in children (see Laplane et al. 2019). On the other hand, the success of Newton's work propelled the philosophical discussion on free will and determinism, functionalism was inspired by the advances in computer science and technology (Appiah 2003, 128), and empirical studies on psychopathy had a direct impact on the moral sentimentalism vs. rationalism debate (see, e.g., Nichols 2002; Prinz 2007).

of medicine like Pellegrino. Exploring the relevant questions about the nature of philosophical inquiry in relation to science helped bring to light not only continuities between philosophical and scientific inquiry, but also special normative competences in philosophical work. The assessment of these objections helped clarify fundamental aspects about the nature of philosophical inquiry in relation to science. The view proposed here regards philosophy and science as continuous forms of inquiry, but agrees that philosophy is a normative discipline in the sense of being a discipline with special expertise in normative aspects. This is in part because philosophical work not only deals with epistemic norms about how we ought to form beliefs or behave as inquirers, but also with the practical sources of epistemic norms that explain the *reasons* why we ought to care about them (Kornblith 1993; 2002).

CHAPTER 3

Science and Medicine
The Systematicity Thesis

3.1 Introduction

Any analysis aiming to address general questions about the nature of medicine needs to consider the scientific status of medicine. Surely, it would seem that medicine is a proper science, firmly grounded in other scientific areas such as biology and chemistry. Large-scale randomized placebo-controlled trials, diagnostic genetic sequencing, radiation therapy, and stem cell therapies appear to testify clearly to the scientific nature of modern medicine. But are such characteristics sufficient for concluding that medicine is a science?

Some deny this, arguing that medicine is in its essence not a science at all. Some acknowledge that medicine *uses* scientific knowledge to support its judgments, but maintain it does not qualify as science, because it "is largely observational and functions without the level of certainty essential to science" (Miller and Miller 2014, 151; Miller 2014). Others argue that medicine is essentially different from the "pure" search for truth and knowledge that characterizes the sciences. Its pursuit of knowledge is "determined by a practical end which truth serves, namely health and healing of human beings" (Pellegrino 1998, 327). Such a view distinguishes between medical science and medical practice, and maintains that philosophy of medicine is concerned with the latter, whereas the former can be reduced to the philosophy of biology and science (Pellegrino and Thomasma 1981, 26–7).[1] Philosophy *of* medicine on this view is reserved for inquiries that are ultimately grounded in the individual doctor-patient relationship. This is seen as distinct from philosophy *in* medicine, which involves the application of philosophical tools to medically defined problems, much like in philosophy of science (Pellegrino 1976; 1986; 1998).

[1] As a consequence, as we have seen in Chapter 2, philosophy of medicine is on this view distinct from philosophy of science (see Pellegrino 1998, 326).

We have rejected this separation in Chapter 2 and offered an account of philosophy of medicine as a subdiscipline of philosophy of science. Nonetheless, we have not directly taken issue with the claim that medicine is not science, which is not only an interesting and challenging one, but one that might be taken to bolster the skeptical challenges sketched in Chapter 1.

For this reason, it is necessary to confront questions about what it is that makes medicine scientific and distinguishes it from nonscientific enterprises. With medicine playing such an important role in almost all domains of contemporary life, addressing questions about the scientific nature of medicine is of obvious urgency. Depending on how these questions are answered, it might well affect our epistemological and moral expectations with respect to medicine (Munson 1981; Cunningham 2015).

The main thesis that this chapter seeks to defend is that medicine is *systematic inquiry,* thus fulfilling a crucial, necessary criterion for science. In order to fulfill this task, we need to comprehend what it is that characterizes scientific inquiry.[2] Long before science entered the scene, human beings engaged in inquiry and attained vast amounts of knowledge and understanding through the exercise of "commonsense" methods. They learned regularities about fire, weather, the seasons, and soil fertility; gained insight into the foods and liquids that best nurtured and hydrated their bodies and the kinds of materials that offered the best shelter; and the list goes on. Compared to such commonsense methods, what defining characteristics do scientific inquiry and knowledge-acquisition display?

To answer this question, we may for now leave aside discussing what inquiry is aimed at. While we will discuss this in detail in the following chapters, for our current purposes we may remain neutral on the question whether inquiry aims at producing true beliefs, knowledge, or understanding. Also, we may remain neutral on the question whether inquiries directed at settling a particular question are fundamentally different in this respect from inquiries directed at general phenomena (for a discussion, see Kelp 2021).

[2] Some tend to emphasize how science has transformed society, not merely by offering knowledge about various types of events and processes that then enabled them to be controlled, but by contributing to realizing ambitions connected to the idea of a liberal society. By putting under scrutiny domains previously shielded from critical thought, science has weakened, as Nagel (1961, vii) aptly puts it, the "intellectual foundations for moral and religious dogmas, with a resultant weakening in the protective cover that the hard crust of unreasoned custom provides for the continuation of social injustices."

This chapter starts by consulting the literature on the "demarcation" problem in the philosophy of science, which deals with the characteristics that distinguish science from nonscience, bad science, or pseudoscience. Learning from failures of earlier approaches while at the same time acknowledging the importance of demarcation, the general view proposed here is that we should adopt a *Deflated Approach*, which denies discipline-independent necessary and sufficient conditions, and acknowledges that "science" is a family resemblance concept that admits differences of degrees to nonscientific undertakings. Then, drawing on Paul Hoyningen-Huene's (2013; 2019) account of systematicity as a necessary criterion for science and Alexander Bird's (2019b) analysis of historical examples showing a correlation between the increase in medicine's scientific character and the systematicity of its methods, this chapter seeks to show that medicine meets the requirement for systematicity on all dimensions. While lacking necessary *and* sufficient conditions systematicity theory cannot fully establish that medicine is science, it nevertheless gives us good reasons to accept this thesis.

The chapter then considers and defuses an objection that other, nonepistemic differences linked to the distinctive duality of medicine warrant thinking that medicine is not science. In the final part, the chapter simultaneously deploys and tests the usefulness of systematicity theory. Expanding work by Holger Lyre (2018) and opposing the view by Naomi Oreskes (2019) that systematicity fails as a criterion of demarcation from pseudoscience, it is shown that homeopathy, widely regarded as a pseudoscience, does not exhibit the type of synchronic and diachronic systematicity that characterizes medicine and scientific endeavors in general.

3.2 Problems with Demarcation

Since Karl Popper portrayed the demarcation problem as the key to a large number of basic issues in the philosophy of science (Popper 1962, 42), a significant number of attempts have been made to offer criteria for demarcation. Popper himself proposed falsifiability, while others have suggested that science and pseudoscience can be distinguished in terms of the former's ability to solve puzzles (Kuhn 1974), or in terms of its being adequately integrated into the other sciences (Reisch 1998). Yet others have argued that while no demarcation can be offered at the level of theories, whole research programs can be assessed by exploring whether new theories developed within their frameworks exhibit a growth in empirical content (Lakatos 1981). These attempts are now considered to have failed, and the prominent view is that the tremendous diversity of methods and theoretical frameworks characterizing different disciplinary traditions makes it impossible to offer discipline-independent necessary and sufficient conditions. Embracing a skeptical conclusion, Laudan (1983) dismisses the demarcation problem as misguided and futile. As he (1983, 123) puts it, "the evident epistemic heterogeneity of the activities and beliefs customarily regarded as scientific should alert us to the probable futility of seeking an epistemic version of a demarcation criterion."

The skeptical conclusion is unsatisfactory, especially when we consider the case of medicine in light of some of the criticism that Chapter 1 portrayed. First, saying that scientific knowledge, for example, in medicine is just one form of knowledge among others is unsatisfactory and does not explain why the former is incomparably more reliable than any other kind of knowledge. Second, and more importantly, the demarcation of scientific knowledge from nonscientific knowledge has a crucial role in liberal democracies, which rely on science as a provider of unbiased facts that can offer guidance in the political realm (Hansson 2017). Many will agree that tax funds ought to support *scientific* medical research rather than nonscientific projects, and that national and supranational public health policymaking, health care, criminal justice, and education ought to be based on properly scientific knowledge. In many cases, energy policies are informed by scientists who model the effects of continuing reliance on fossil fuels, educational policies are informed by findings about the relationship between learning environments and learning outcomes, and medical devices are approved on the basis of scientific scrutiny of their efficacy. Third, demarcation is also vital because pseudoscience is now no longer a relatively harmless leisure activity of a relatively small number of individuals. By feeding on the growing distrust of established authorities and by mimicking science, pseudoscience has grown to a size that has genuine impact on societies. For example, creationism has changed the course of public education in a number of places, alternative medicine has offered false hope for numerous patients, pseudoscientific theories about vaccines and AIDS have resulted in lives lost, and sects based on pseudoscientific beliefs continue to cause damage in human lives (Pigliucci and Boudry 2013). Fourth, questions about the distinctive status of science as a source of knowledge are pressing in times of declining public trust in science. Progress in this area is poised to make a significant societal impact, while public ignorance of the difference between science and what pretends to be science is damaging. It emboldens the populist view according to which the epistemic authority of science derives solely from institutions and social processes, not from the nature of the body of beliefs (Dupré

1993). On this view, "scientific" is nothing but an elitist label that is not grounded in features that afford scientific knowledge a higher status than knowledge originating from other sources.

3.3 The Deflated Approach

Demarcation thus appears both necessary and unachievable, as scientific inquiries do not consistently share methodology or subject matter. A potential solution could proceed along Laudan's thoughts on this matter. Laudan (1983, 123) distinguishes between two questions, and argues that the first one ("What makes a belief well-grounded?") is stimulating and manageable, while the second one ("What makes a belief scientific?") is not. While Laudan thinks that terms like "pseudoscience" and "unscientific" should be eliminated because they only express attitudes, he also stresses the importance of preserving the distinction between reliable and unreliable knowledge.

It seems correct to assume that this is a core issue. However, the two questions should not be separated. After all, settling the question of reliable knowledge and the methods that lead to it will include exploring the class of statements commonly considered as falling under the label "science" in counter-distinction to the class of statements falling under the label "pseudoscience." Also, since commonsense methods can produce reliable knowledge, it seems that we need a special rubric for the type of reliable knowledge that we need to inform our policies that goes beyond commonsense knowledge.

A potential solution is an approach – call this the Deflated Approach – that does not abandon attempts at offering general criteria, but only the idea that these take the form of an ahistorical, discipline-independent set of individually necessary and jointly sufficient conditions. Given the lack of commonalities that characterize the entire range of activities or systems of belief that coexist under its conceptual umbrella, "science" ought to be comprehended as a family resemblance concept. It contains a large class of items connected by family resemblance only, without recourse to criteria that apply to each and every item in the whole class. Thus, rather than looking for necessary and sufficient criteria for what qualifies as science, it is more suitable to seek to offer a cluster of standards in the form of epistemic virtues. As John Dupré (1993; 2004, 26) acknowledges, "the best we can do is to draw up a list of epistemic virtues and apportion our enthusiasm for knowledge-claiming practices to the extent that they meet as many as possible of such criteria."

The *Deflated Approach* can draw inspiration from work on epistemic vices in pseudosciences. These seem to be characterized by overly strong trust in authority, a lack of inclination to test hypotheses, indifference to contradicting data, cherry-picked examples and testing procedures that confirm a theory, abandoned explanations without replacements, or replacements with new theories that increase the number of unexplained aspects (see, e.g., Hansson 2017). Importantly, these vices are taken to define pseudoscience without amounting to necessary and sufficient conditions. Shifting to the virtues of science, Thomas Kuhn (1977, 321–2) provided initial impetus by offering a list of virtues for assessing scientific theories, which includes accuracy, consistency, broad scope, fruitfulness, and simplicity. The list could be expanded to include predictive accuracy, internal coherence of theory components, consistency with prior well-established theories, testability, unifying disparate parts under a few primary hypotheses, and fertility (McMullin 1983; Ruse 2009). While no single epistemic virtue is necessary or sufficient, whether a theory or activity qualifies as science will depend on the extent to which it embraces these epistemic virtues.

The demarcation of science with the help of the Deflated Approach leads to two sorts of continuity. First, whether a theory or activity qualifies as science is a matter of degree. What characterizes science is not some intrinsic property, but a *relational property* that admits differences of degrees, not categorical differences. Second, because these virtues to some extent also characterize nonscientific undertakings, the Deflated Approach acknowledges a continuity between scientific and nonscientific practices of knowledge production that exhibit overlapping epistemic virtues.

The advantages of such continuity notwithstanding, some might raise doubts about whether the approach is too permissive to offer suitable criteria for demarcation. To avoid this problem, the Deflated Approach needs to identify and explicate an overarching epistemic virtue that gives enough unity to the list of epistemic virtues while showing how it can be used to distinguish (a) knowledge acquisition in science from knowledge acquisition by commonsense methods, (b) scientific knowledge from commonsense knowledge, and (c) science and pseudoscience. A promising way to achieve this is to draw on the notion of *systematicity*.

3.4 Incompleteness and Systematicity

It is little disputed that inquiries using commonsense methods can yield knowledge and understanding. Human beings living in early agricultural societies came to know that spreading manure preserves the fertility of the soil. In a long process that also involved forms of experimentation, which is in many ways continuous with science, they also came to know countless other details required for successful agriculture, such as the appropriate amount of manure and the right time of the year to spread it. Nonetheless, such commonsense knowledge has its limitations, in part because it is often not accompanied by an awareness of the limits of its validity. In spite of the manifest deterioration of the soil, people continued to spread manure and were often helpless in the face of a critical problem of food supply. Commonsense knowledge about manure as a fertilizer proved particularly vulnerable to changes because it was incomplete, lacking comprehension of the reasons for its successful operation. The effectiveness of manure depends on the persistence of conditions of which common sense is usually unaware and which can only be understood in light of principles of biology and soil chemistry.

In contrast to commonsense knowledge, the knowledge produced by agricultural science is far more complete and refines ordinary conceptions by unveiling hidden conditions. The difference, as Ernest Nagel (1961, 5) has noted, can be explicated in terms of an increase in *systematicity*. This permits a gradual progression from prescientific (or nonscientific) to scientific knowledge and grants scientific knowledge a higher degree of systematicity without rendering the commonsense knowledge unsystematic.

At a minimum, systematicity implies some kind of order and absence of arbitrariness, but its more concrete meaning is best unfolded and made explicit in the contexts in which it is used. Paul Hoyningen-Huene (2013) has examined and elaborated this notion, drawing on the German tradition, which comprehends science as "Wissenschaft." He argues that there are at least nine dimensions along which science displays more systematicity than other kinds of knowledge. This leads to nine distinct theses, claiming higher systematicity for each of those dimensions of science (i.e., descriptions, explanations, predictions, the defense of knowledge claims, critical discourse, epistemic connectedness, an ideal of completeness, knowledge generation, and the representation of knowledge). The concepts of systematicity corresponding to the nine dimensions are descendants of an abstract concept of systematicity, but they are specified in a context-dependent manner. For this reason, instead of necessary and sufficient conditions, different forms of systematicity are held together by family resemblance relations. For example, what qualifies as a systematic *description* will vary between disciplines, and what qualifies as a *systematic* description within a discipline will be different from what qualifies as a systematic explanation in the same discipline.

The dimensions exhibit some overlap and have some limitations in their applicability. Not all dimensions apply to all disciplines (e.g., historical natural sciences like paleontology make no predictions), and in some disciplines certain dimensions may be used in divergent ways (e.g., history operates with descriptions that might elsewhere qualify as explanations). The unity that systematicity lends to the sciences is a unity based on relations of family resemblance: sciences share characteristics with their neighboring sciences, without there being a single characteristic that is common to all of them. They belong to the same class of enterprises that generate knowledge in a way that is more systematic than other enterprises.

The following sections examine the evidence that scientific knowledge is more systematic than its everyday equivalent, while at the same time focusing on systematicity in medicine. In support of the *Systematicity Thesis*, it is argued that medicine displays systematicity along all nine dimensions.

3.4.1 Description

The description of the subjects of scientific endeavors typically involves procedures like categorization and classification that tend to increase systematicity and reduce the diversity of individual items. In this process, qualitative descriptions (e.g., "water is freezing cold") are replaced by quantified descriptions (e.g., "water is 2 degrees"), increasing interrater reproducibility and precision: a thermometer allows distinguishing hundreds of different temperature states, while qualitative everyday language offers only few descriptions. The increase in quantitative accuracy in determining temperature leads to a richer system of description. Some sciences (physics, chemistry) aim at generalized descriptions of classes of events or processes and focus on regularities holding in the respective domain, abstracting away from individual features. This is often the case in political science or sociology, where one such well-known empirical regularity seems to be that democracies are not at war with each other. Other sciences describe individual phenomena. For example, historians use narrative form to describe a sequence of events, situations, and processes that are relevant to the subject in question.

In medicine, description involves classification procedures, which categorize and help reduce the diversity of individual items that increase systematicity. Efforts at classification greatly increased in the wake of the great pandemics of plague in the fifteenth century. The collection of data from death certificates was increasingly used to monitor epidemics, and this constituted the origin of the *Bertillon Classification of Causes of Death*, which served as the basis for the first international classification in 1893. The 1893 classification went on to become the *International List of Causes of Death* and ultimately the World Health Organization's (WHO) *International Classification of Diseases* (ICD), which is now an internationally recognized tool that provides a shared terminology for registering and monitoring purposes.

Importantly for our purposes, the ICD is a good example of the kind of systematic description that characterizes scientific endeavors. First, the most recent version in use, ICD-11, not only contains medical terminology for effective communication about disease, description of diseases, and their diagnostic characteristics, but also a unique, internationally recognized identifier that enables the collection of data for purposes of analyzing trends and changes in mortality and morbidity.[3] The ICD uses a code that combines letters and numbers, from A00.0 to Z99.9, in such a way that they simultaneously designate a chapter (out of a total of 28) and specify different classification axes. Since World War II, when the United Nations entrusted responsibility for the ICD to the World Health Organization (WHO), the ICD has been periodically revised (in 1948, 1955, 1965, 1975, 1989, 2018) with each revision increasing in precision and detail, which enables intensified international data collecting for purposes of policymaking and epidemiological surveillance. To increase transparency, the WHO also publishes official annual updates, which list changes. For example, a recent addition is a new classification system for chronic pain.

Such descriptions and classifications display crucial differences to those used in everyday life. Our everyday classifications are limited and typically closely connected to practical matters; we do not replace qualitative descriptions by quantified ones, and when we describe by dividing processes into phases and using a narrative form, the degree of systematicity never reaches that of psychology, sociology, or history. It does not abide by principles of historical continuity, and the elements selected for the story are not strictly guided by relevance.

[3] The ICD-11 came into effect in 2022.

3.4.2 *Explanations*

Explanations in everyday life and science share a number of common features. Everyday explanation sometimes involves theories about unobservable entities (e.g., the mayor of this city does not believe that road maintenance is important), but these are informal compared to their scientific counterparts, which posit theoretical entities that are not directly observable (e.g., microinflammation in the brain, electrons, gravitational force in Newton's gravitational theory), or unobservable by their nature (e.g., ideal types like *homo economicus*). Both refer to some type of regularity (e.g., the traffic light is broken because this city has no budget for maintenance), but in the case of everyday explanations, these are typically not explicit and detailed and are much more prone to bias than their scientific counterparts, some of which make recourse to laws of nature. To uncover some of the details, it is useful to examine the difference in causal explanations in both everyday and scientific contexts.

For instance, explaining depression might involve identifying a proximate cause (certain imbalances in brain chemistry or environmental precipitants). Everyday causal explanations often take on a minimal form, explaining event X as the cause of event Y if X precedes Y and if X is linked to Y in a consistent manner. However, such minimal forms do not offer an ontological basis for the association and fail to capture how causal explanations in many cases allow scientists to manipulate phenomena on the basis of explanations. As we shall see in more detail in Chapter 5, causal explanations in the sciences are richer, for example, in the sense that they specify the cause as necessary (the absence of X guarantees the absence of Y), sufficient (the presence of X guarantees the presence of Y), or both. Moreover, they refer to actual causal mechanisms and explicate the interaction of the parts of those mechanisms. For example, explaining depression might involve identifying an ultimate cause, which could be given by exploring whether depression has an evolutionary function. Ultimate (evolutionary) causes offer a broad picture, as they explain a process or a structure in all members of a species, while proximate causes explicate processes or structures in individual organisms.[4]

[4] Theories from evolutionary medicine offer ultimate-cause explanations for why we have appendixes, wisdom teeth, and narrow birth canals, and why we suffer from sickle cell disease and some of the most common mental disorders. Because mental disorders occur at prevalence rates that are too high to be explained by mutations, they likely have a genetic basis that has been promoted by natural selection. Three types of evolutionary explanations can be distinguished (Murphy 2005; Varga 2015). *Breakdown explanations* understand mental disorder as the malfunction of some mental or neural component in fulfilling its evolutionary function. *Mismatch explanations* understand mental

Both everyday and scientific explanation can be reductionistic. Many everyday cases of explanation involve the components of objects (e.g., the machine in the kitchen does not brew coffee because the water container is blocked), but they do not display the kind of complexity that, for example, reductionist explanations in medicine do. Medicine offers bottom-up explanations of the functioning of an organism that draw on the functioning of its organs or other constitutive parts, and the functioning of these parts by recourse to lower-level parts (e.g., blood vessels, cells, etc.). There are a number of options for reductive explanation, the choice of which is in large part determined by interests in prevention and control. For example, the cause of a person's death can be described in a number of ways. With reference to the same person, a physician might attribute the cause to lung cancer, the coroner to pulmonary embolism, or the epidemiologist to smoking. Often, the adequacy of explanation depends on the context and the ability to intervene. Surely, the causal history of this person's death may have included genetic mutation and social pressure (to smoke), but these are often not contained in explanations, because, at least given the current state of things, they do not specify a proper target for medical intervention.

Both everyday and scientific explanation uses "inference to the best explanation," which roughly amounts to choosing the hypothesis that explains a phenomenon in a more satisfactory manner than rival hypotheses. A famous illustration that shows the difference to everyday cases is Ignaz Semmelweis's explanation of the difference in mortality from childbed fever between two maternity wards located in the same hospital. Having considered and rejected hypotheses based on psychological factors, overcrowding, diet, and expert care, Semmelweis observed that medical students cared for patients in the high mortality ward. Excluding the possibility that the increased mortality rate was linked to the way in which medical students conducted examinations, Semmelweis judged that the hypothesis that provided the best overall explanation was that medical students were infecting patients with "cadaveric matter," even though such material was not observable (Lipton 2004).[5]

disorder as connected to mechanisms that were once adaptive but no longer match our modern environment. *Persistence explanations* hold that some disorders qualify as adaptive in the present environment.

[5] Some explanations in the social sciences and the humanities do not refer to theories, but to intentions and beliefs. To explain why Obama signed the Affordable Care Act into law, we make recourse to Obama's political intentions to make health care coverage available to more people and his belief that this act would realize his intention. Explanation by ascribing particular intentions is a common pattern in everyday life, but in that context, the ascription of intentions is often done without much supporting argument. Historical explanations often consist of narratives that clarify

3.4.3 *Predictions*

While not all scientific endeavors engage in predictions of not-yet-observed events, those that do typically draw on regularities discovered in their empirical data. Predictions can be based on theory, as was the case with the 1846 discovery of Neptune. On the basis of discrepancies between calculations based on Newton's theory of gravitation and data for the orbit of Uranus, scientists were able to predict the position of a yet undiscovered planet that gravitationally influenced Uranus's orbit. There are of course many similar cases of predictions based on theories, but many predictions of the behavior of complex systems are today based on models and simulations. For example, in meteorology, a model of Earth's atmosphere consists of a set of points, a set of variables (describing the actual weather state, e.g., pressure, temperature, wind, water vapor, etc.), and a set of dynamic equations involving these variables. The equations simulate the change of the weather system, such that the initial weather state at each grid point at some time and some initial conditions (e.g., surface conditions) allows predictions about the future weather.

Prediction is a central concern in epidemiology, health policy, public health, and clinical practice, and exploiting a range of empirical data for predictive purposes constitutes a central activity in medicine. On the individual level, being able to predict the probability of developing disease might help prevent it or significantly decrease its impact. While biological markers usually identify a disease process that has already started, genetic testing allows estimating the risk of developing a disease. On the population level, epidemiologists study the distribution and determinants of diseases like obesity and make predictions on, for instance, the effect of reducing obesity in a population with respect to mortality. Public health institutions that design interventions to reduce disease transmission rely on mathematical models of infectious disease dynamics. Using such models and tracking patterns in public health reports, it is possible to identify signs of an impending epidemic in the general population on the basis of relatively subtle changes (Drake et al. 2019).

In everyday life, we use diagrams and models to visually illustrate relationships between entities and events, and make predictions based on

the causal chain leading to an event, without recourse to some general theory that would connect them. For example, to explain the emergence of welfare states after World War II, certain unique events (e.g., strikes) must be linked to more general tendencies (e.g., the economic situation in European countries). Although they resemble the structure of some everyday explanatory stories, historians carefully document their claims and exclude as far as possible alternative explanations.

simple models. These contain simplifying assumptions and small-scale models of reality, but they are less detailed and systematic than their scientific counterparts. Moreover, everyday models and predictions lack the elaborate data, calibrated models, and systematic description on which predictions in science are based.

3.4.4 The Defense of Knowledge Claims and Critical Discourse

In our everyday lives, we aim to remove error in our pursuit of knowledge, and we are also often prepared to offer defense of our knowledge claims. However, scientific inquiries are much more meticulous and effective in identifying and removing sources of error. In the formal sciences, a thorough way to eliminate error is to provide a proof, while in the empirical sciences observational or experimental data play a dominant role. Empirical generalizations, models, and theories differ with regard to the procedures they have in place to minimize error, but they typically offer hypotheses and predictions that allow confronting theoretical constructs with empirical data.

To avoid errors and mitigate the effects of various forms of bias, claims in medicine about causal factors are typically modest. For example, regularly observing that symptom X (fever, fatigue) is followed by symptom Y (weight loss and chronic diarrhea) is not sufficient for establishing a causal connection between X and Y. Rather, being infected with a virus Z can cause both, without there being a causal relation between X and Y. In cases in which the host's genetic makeup is suspected to have a causal role in disease progression and susceptibility to infection, causal claims are normally only made on the basis of studies that investigate the questions by using animal models that lack a particular gene. Because the assessment of a medical intervention is especially difficult, medical research adheres to procedures that aim to prevent bias. For example, *confirmation bias* occurs when scientists focus on data that support their hypotheses and stop gathering data when the retrieved evidence confirms their theory (or they disregard other data entirely). To avoid such bias, among other measures, scientists acknowledge the results of all available studies, follow rigorous protocols for executing the research, and specify a number of criteria prior to embarking on the project.

The way in which science is organized shapes the way in which knowledge generated by scientific practice is handled. At least in most cases, scientists are socialized into communities with a set of norms that opposes bias and fraud and institutions with high epistemological standards that are beneficial for the success of the enterprise. To maintain these standards, there is a general requirement that knowledge claims be thoroughly defended. Knowledge claims are examined by scientists in a number of ways, and community members participate in the critical dialogue about the generation and adjustment of standards by which theories, hypotheses, findings, and observational practices are assessed. The communication of results is organized to help direct attention to minimizing errors and eliminating bias and fraud. For example, researchers submit the results of their studies to peer-reviewed journals along with the raw empirical material they have gathered. In the review process, which is itself "blind" or "double-blind" to avoid bias, knowledge claims are scrutinized, and authors receive comments and criticism that help increase the quality of the scientific work. The presentation of findings at conferences is followed by a Q&A period, which offers a chance to consider objections and fosters critical discourse.

We should be careful not to paint an overly optimistic picture with respect to fraud and misconduct and the ability of institutions to prevent it. In fact, some recent discussion and controversies underline serious shortcomings with implementing secure and reliable measures in review and publication procedures (e.g., Boetto et al. 2021). One solution to secure improvement is to submit not merely the creation, collection, and sharing of research data to review, but also the editorial process itself. Blockchain technology is in this regard often suggested to ensure traceability. But there is lots of room for improvement with respect to measures already in use. For example, while most journals now demand the submission of data-sharing plans and statements, many of them do not explicitly state which kinds of data-sharing plans they accept.

3.4.5 Epistemic Connectedness and the Ideal of Completeness

Epistemic connectedness refers to the display of robust links to other pieces of scientific knowledge, which nonscientific areas of knowledge production by professionals typically lack. For example, chess theory builds on mathematical game theory, yet it does not count as part of mathematics, in part because its content is not connected to other mathematical areas. Alternatively, compare an article on arthritis published in a newspaper with a paper on the same subject in a professional, peer-reviewed philosophy journal. Both authors might have read the same literature, but one major difference will be that the paper in the philosophy journal will have a bibliography and footnotes that show a large number of links to other sources of scientific knowledge. Moreover, the journal article will explicitly state its thesis, identify background theories and their sources, and examine how the thesis of the paper challenges or confirms other theories. It will have an abstract and a list of keywords that facilitate making connections and it will place itself into the existing landscape of theories and argue that its content should be taken as a novel contribution. The newspaper article may be of the highest journalistic quality, but it will not display the same degree of epistemic connectedness as the journal article. In the same way, medicine seeks to systematically connect with relevant disciplines like biology, psychology, and physics, leading to a number of subdisciplines like evolutionary medicine, psychoneuroimmunology, and medical physics.

Moreover, in contrast to everyday knowledge, scientific disciplines do not content themselves with scattered facts about a certain domain, but strive to expand and complete their knowledge. In the case of scientific classifications and taxonomies, completeness is not merely an implicit ideal, but an explicitly articulated goal. Chemists seek to provide a comprehensive system of elements, biologists aim at a complete description of all biological species, and medical researchers work toward attaining a complete taxonomy of diseases. Great examples that embody the ideal of completeness are the periodic table, which took more than fifty years to complete, and the human genome project that aimed to chart the genes in human DNA, which took thirteen years. Meticulous reference works like *encyclopedias* provide summarized knowledge from a range of fields and disciplines. For example, the 2010 version of the general knowledge *Encyclopedia Britannica* consists of 32 volumes and has 100 editors and more than 4,000 contributors. It clearly follows an ideal of completeness: articles are updated at regular intervals, and since it was first published during the late eighteenth century, it has grown by a staggering 29 volumes.

3.4.6 Knowledge Generation and Knowledge Representation

Unlike everyday knowledge-seeking practices, science is constantly striving to attain new information by performing new observations, changing parameters in extant experiments, and improving observational and experimental tools. But science is also constantly improving the quality and accessibility of existing data, for instance, when deploying new procedures to systematically search archives (by exploiting state of the art information technology) or new technologies to increase knowledge about the age and composition of historical artifacts. In a self-amplifying process, existing knowledge is used to generate new knowledge, which provides additional resources for further knowledge expansion, increasing the number of subspecialties, researchers, scientific journals, publications, and so on.

New technological developments enable multipronged investigations that produce new data. For example, to find evidence of historical changes in climate climate scientists combine written evidence of climatic variations, pictorial representations from nonliterate societies, and data about population levels of humans, plants, and animals. These can be brought together with findings from paleoclimatology, which offers data about the climate prior to the availability of records through analyzing ice cores from continental glaciers from both hemispheres. A good example from medicine is the use of ICD codes to capture huge amounts of data through the routine operation of medical care. ICD codes are not only used to serve the objectives for which they were originally devised, they now enable analyzing patterns of disease distribution, health outcomes, and standards of care. At the same time, with additional resources, mapping disease trends based on ICD codes not only assists comprehending the general health of a population, but allows inferences about the effectiveness of care, social determinants of health (e.g., nutrition, environment, education), and public infrastructure.

Due to the large amount of information generated in the sciences, knowledge representation techniques were developed that are themselves systematic. By using visual techniques and specific formats, knowledge is represented in a way that supports quick and effective transmission. In medicine, as in other sciences, vast amounts of knowledge are made available for reliable decision-making and rendered as transparent as possible to help in detecting errors and missing pieces.

3.5 Medicine and Systematicity

In the previous sections, systematicity was introduced as an umbrella concept that covers connections of family resemblance between scientific disciplines. The nine dimensions include features that distinguish their bearers only in degree from their nonscientific counterparts. We have examined evidence for the *synchronic claim* that knowledge produced in the sciences and in medicine is more systematic than other kinds of knowledge, particularly its everyday counterpart. But the material also offers reasons for accepting the *diachronic claim* that the development of a scientific discipline is accompanied by an overall increase in systematicity. Applied to the case of medicine, we have seen that compared to everyday knowledge, an increase in systematicity occurred in description procedures (introduction of new classifications), explanation (combination of several types of causal explanation), prediction (introduction of new mathematical models), defense of knowledge claims (new bias elimination methods, evidence-based medicine [EBM]), knowledge generation and representation (using AI to analyze ICD codes), epistemic connectedness (evidenced by new subfields), and the ideal of completeness. On such a basis, it seems safe to conclude that medicine is systematic both synchronically and diachronically and thus fulfills a necessary condition for science.

Consider, for instance, the dimension of defense of knowledge claims. In a much more detailed fashion than in this chapter, Alexander Bird (2019b) focuses on this dimension and uses examples from medicine in the eighteenth century to show a correlation between the increase in medicine's scientific character and the increase in the systematicity of its methods, which effectively minimize cognitive bias (e.g., Jurin's assessment of the safety of variolation). On this basis, Bird motivates a closely related hypothesis, according to which systematicity is characteristic of science because it allows overcoming the limitations of ordinary cognitive capacities that succumb to various forms of biases. Introducing systematicity allows reliable reasoning processes, which lead to the acquisition of knowledge that we would be unable to obtain if we merely applied our everyday cognitive capacities. The beliefs acquired by using everyday cognitive capacities are often not reliably formed, are inefficient in excluding biases, and hence fail to constitute knowledge.[6] In contrast, the reasoning processes leading to accepting propositions distinctive of science are reliable and knowledge-generating to the extent that they are systematic.

One might grant that medicine displays systematicity along all nine dimensions, but argue that this does not amount to much. Because systematicity is a necessary condition and not a sufficient one, it might be very inclusive, and might thus not allow demarcating medicine from nonscience, and more importantly, pseudoscience. We have said that lacking demarcation is not just a theoretical problem in philosophy but also very much a practical problem, aggravated by the increasing prevalence of pseudoscientific information that impacts public discussions about policy. For this reason, the aim of the following sections is to show that systematicity can be put to

[6] Hoyningen-Huene (2013, 21) uses "knowledge" in the sense that he thinks scientists use it, namely as a well-established set of beliefs widely accepted in the scientific community. Bird (2019b, 865–6) argues that the link between systematicity and knowledge is that science aims at reliably produced true belief, to which systematicity is conducive. The position defended in this book ultimately deviates from both. As it will be argued in Chapter 4, the aim of science is not knowledge, but understanding.

use to distinguish medicine from homeopathy, which, for our current purposes, we will assume is a pseudoscience. This seems unproblematic, as homeopathy often appears in discussions about the demarcation of science from pseudoscience and is taken to be one of the clearest examples of pseudoscience (Pigliucci 2015; Mukerji and Ernst 2022).

3.6 Systematicity and Demarcation

The demarcation attempt here will not evaluate intrinsic properties of a scientific field, like Popper did in his work, but proceed instead in a doubly comparative fashion.[7] It will test *synchronically*, by comparing systematicity of a reference science and a putative pseudoscience at a given time, and *diachronically*, by comparing the increase in systematicity of a reference science and putative pseudoscience over a certain time span.[8] If a putative pseudoscience fails to exhibit diachronic and synchronic systematicity compared to the reference science while claiming to be science, then the label pseudoscience is warranted.

For our purposes, based on Lyre's (2018) initial proposal on this matter, medicine will be compared with a candidate purported pseudoscience, homeopathy. A thorough examination of the homeopathy literature is beyond the scope of this discussion and the aim is not to assess whether homeopathy is effective, but whether systematicity as a demarcation criterion would exclude homeopathy from being scientific, which would be consistent with what scientists and physicians overwhelmingly think. Thus, we proceed by assuming that it is a pseudoscience and test whether demarcation deploying synchronic and diachronic systematicity criteria reaches the same conclusion. If the answer is in the positive, then we have further reasons to think that systematicity is a *necessary criterion* for science.

Present-day homeopathic practice is based on principles that have remained basically unchanged since they were articulated in the early nineteenth century by the German physician Samuel Hahnemann (Smith 2012). Dissatisfied with invasive and unsanitary medical interventions (e.g., bloodletting, purging, the use of mercury), Hahnemann's experiments on himself and some followers led to three treatises, which

[7] Pseudosciences typically compete with established sciences (creation science competes with evolutionary biology), but there might be cases in which no competitor exists for a given putative pseudoscience.

[8] Hoyningen-Huene (2013, 203–4) rightly argues that, in many cases, the most adequate approach is to compare progressive trajectories over a longer time span. Nonetheless, if possible, which I think it is here, focusing on both synchronic and diachronic aspects provides a richer picture.

posit that healing is a matter of stimulating a "vital force" and identify two basic principles (the "law of similars" and the "law of infinitesimals") (see Bellavite et al. 2005). Let us take a closer look at these principles.

According to the *law of similars*, a substance that gives rise to certain symptoms in healthy people can be used against a disease characterized by the same symptoms. Hahnemann found that while quinine alleviates the symptoms of malaria, ingesting quinine leads to the development of symptoms common to malaria. Together with some followers, Hahnemann conducted tests in which healthy people ingested various substances while the subsequent symptoms were observed and recorded. Conducted long before any methods for clinical trials had been laid down, the procedures were liable to selection bias (differences between baseline characteristics of the compared groups). Without double-blinding, experimenters and participants were knowledgeable about which intervention was received, and this, rather than the intervention itself, might have affected both outcome (performance bias) and outcome measurement (detection bias). Participants were not compared to controls, and significant variations occurred in the quantity of the administered substance and the length of the observation. Many of the symptoms recorded were highly subjective and may thus have been biased by suggestibility. Overall, the studies do not permit inferring a causal link between symptoms and the administered substances. The lack of systematicity means that the studies are not protected from various kinds of bias, and beliefs based on them would not be reliably formed.

Defenders of homeopathy frequently highlight apparent victories during epidemic outbreaks, thus "the distinct superiority of homeopathy in treating the various epidemics of typhoid fever, cholera and yellow fever which raged across Europe and America in the 1800s" (Bellavite et al. 2005, 447). The 1854 cholera epidemic in London is often mentioned in this regard, because data indicate that patients who received conventional treatment at the Middlesex Hospital had a survival rate of 47 percent, which pales in comparison to the 84 percent survival rate for those who received treatment at London Homoeopathic Hospital. This is taken as evidence that homeopathic remedies do better than conventional medicine. Nonetheless, it is difficult to conclude much on the basis of the numbers, and the higher success rate could be explained by a number of other factors. First, it is quite likely that The London Homoeopathic Hospital had better hygiene than other hospitals. Second, it likely attracted patients who were wealthier and had access to better food and living conditions. Third, instead of indicating the success of homeopathy, the survival rate at the London Homoeopathic Hospital could be explained not by the fact that the patients received homeopathic treatment, but by the fact that they did not receive conventional medicine, which at that point was probably more harmful than beneficial.

According to the *law of infinitesimals,* extreme dilution over several stages of substances accompanied by a special form of shaking can increase the potency and effectiveness of a substance. For example, to reach the common strength of 30C, the original ingredient has been diluted 30 times by a factor of 100 each time (Ernst and Singh 2008, 98), which means that the substance is almost certain to contain no active ingredient. While this is in stark contrast with the standard view that physiological reactions depend on dosage, it is sometimes argued that the diluent (usually water or alcohol) preserves some "molecular memory" of the original substance. But accepted laws and mechanisms in chemistry or physics are not consistent with an idea of a diluent having a "memory" (or that shaking during dilution can increase potency). If water could carry such information, then drinking a glass of water should lead to powerful effects on the body, because it can be assumed that it carries the memory of bacteria, toxins, and a wide range of other substances that it once encountered. In all, given what we know about the world, it is not the sort of intervention that we would expect to be effective (no known mechanism would allow the dilutions to have an effect on the human body), and there is no evidence that it is effective beyond placebo effects, aside from isolated reports that most often have serious design and execution flaws.[9]

Overall, compared to the reference science, homeopathy fails to exhibit diachronic and synchronic systematicity. Synchronically, the fundamental "laws" of homeopathy run counter to principles in established sciences, and epistemic connectedness to adjacent disciplines is missing. Homeopathic theories are not examined in relation to biology, immunology, or physics. The currently held beliefs in fundamental "laws" are based on processes of belief acquisition that violate the requirements of systematicity regarding the defense of knowledge claims and critical discourse. As to explanations, the conclusions from the London case stand in

[9] A group of independent experts for the Australian National Health and Medicine Research Council (NHMRC) conducted, in 2015, a systematic review summarizing the evidence from 57 systematic reviews regarding the effectiveness of homeopathy for more than 60 clinical conditions. The conclusion was that "there are no health conditions for which there is reliable evidence that homeopathy is effective." More recently, an NHS England document (2017) reviewed eight further systematic reviews published subsequent to the NHMRC review and reached the conclusion that there is no robust evidence of clinical effectiveness.

stark contrast to Semmelweis's systematic investigation and explanation. It is hard to identify any sustained attempt at attaining epistemic connectedness by establishing links to other pieces of scientific knowledge. As to the ideal of completeness and the generation of new knowledge, we do not find much evidence that homeopathy is engaged in a sustained attempt to expand and complete its knowledge.

Diachronically, homeopathy exhibits no or substantially less increase in systematicity than the reference science, and if some of its basic assumptions are tested, this is typically done by outside scientists from the respective reference science.[10] While proponents of homeopathy often mount a defense against criticism that challenges the legitimacy of theoretical or empirical aspects (e.g., by offering auxiliary hypotheses to shield their theories), they do not make new predictions and they offer no novel defenses of knowledge claims with new data. Moreover, they usually do not attempt to expand the scope of application of their theories, to improve explanatory accuracy, or to increase the rigor of the claims.

3.7 Limitations

Diachronic demarcation has some straightforward limitations, mainly because it does not say much about the intrinsic qualities of the field, but remains comparative. First, the comparative nature of systematicity introduces some limitations, because it can only deal with pseudosciences that have scientific counterparts. So, if homeopathy did not have a counterpart that is acknowledged to be genuinely scientific, we could not use systematicity theory to assess its status as pseudoscience. Using ayurvedic medicine as a counterpart for comparison would surely yield a different result.

Second, the lack of independent criteria for determining what counts as the right time span for comparing systematicity increase (diachronic) might render the account vacuous. For example, imagine that over a period of, say, five decades, mainstream medicine does not receive any public or private funding and, as a consequence, progress and increase in systematicity is close to zero. At the same time, imagine that homeopathy demonstrates exactly the same increase in systematicity as medicine. While

[10] Without new indications from basic research, spending time and effort to evaluate homeopathy in large clinical trials is difficult to defend. In a case in which the evidence base is lacking, the likelihood of obtaining patient benefit is low and the costs of such trials would be ethically unacceptable, because they divert funds from more plausible trials.

Hoyningen-Huene thinks that the comparison of overall systematicity increases is relevant for the identification of pseudosciences, such cases show that diachronic demarcation will fail, which is why it is best to approach the matter both diachronically and synchronically.

Naomi Oreskes (2019) acknowledges that systematicity characterizes science, but argues that it fails as a criterion of demarcation from pseudoscience. Using homeopathy as one of her examples, she makes two points. First, she maintains that homeopathy is actually systematic. According to Oreskes, homeopathy has in fact generated a systematic body of work: Hahnemann's studies present substantial quantities of empirical data that were collected in a way that was consistent with the methodological standards of the time. Since then, followers have developed Hahnemann's concepts and arguments, offering descriptions and explanations, striving for an ideal of completeness. There is no reason to think that practitioners of homeopathy are not engaged in critical discourse, as evidenced by thousands of studies published in peer-reviewed homeopathic journals. On such a basis, Oreskes (2019, 889) concludes that "homeopathy is systematic, and given the resources available to it, it appears to have achieved a degree of systematicity that is comparable to that of mainstream medicine."

Second, because homeopathy is both a pseudoscience and displays systematicity, demarcating science from pseudoscience must appeal to some criteria other than systematicity. Instead, Oreskes lists seven criteria, according to which pseudosciences (1) ignore evidence obtained by means of established scientific methods, (2) repeat claims exposed as false or obsolete, (3) misrepresent scientific data, theories, and models, (4) rely on measures (e.g., petitions) that lie outside the norms of scientific procedure, (5) prioritize ideological and economic considerations over epistemic ones, (6) motivate ideological and economic considerations that trump epistemic ones, and (7) do not acknowledge evidence as judged by experts in the relevant scientific fields.

Two comments are due here. First, the considerations in the previous section offer reasons for disagreeing with this conclusion. Homeopathy appears to lack systematicity on most of the dimensions. Second, and more importantly for our purposes, it seems that many of the seven criteria listed by Oreskes are actually covered in the different dimensions of systematicity. In fact, we might say that handling evidence as described in (1) and (7) is covered, as it is incompatible with several dimensions of systematicity. As to the rest, repeating incorrect claims, misrepresenting data, and prioritizing economical/ideological considerations amount to negligent or fraud science rather than pseudoscience. But, even so, the requirements of systematicity and the methods for avoiding biases also offer some protection against fraud involving the intentional fabrication of data and records. Fraud, at least significant amounts of it, is something that methods safeguarding systematicity through the epistemic connectedness and the defense of knowledge claims will often be able to catch.

3.8 The Science versus Scientific Argument

Having explored how the systematic nature of scientific knowledge distinguishes it from other forms of knowledge, we concluded that medicine is systematic and thus fulfills a necessary criterion for science. This demarcation was epistemic with reference to systematicity as a certain type of epistemic characteristic that is conducive to reliable knowledge. Because we have not provided sufficient criteria for an activity to count as science, we need to consider an extra set of objections. Our objector might grant that medicine is systematic, but maintain that there are other, nonepistemic differences that warrant thinking that medicine is not science. More precisely, the objector might stress that the analysis has reached this conclusion by neglecting the distinctive *duality of medicine*, which is why it cannot be a science.

In contemporary dictionaries of medicine, we find that medicine is characterized by the duality of *medical science* and *medical practice*, the latter of which consists of caring for individual patients and their specific conditions.[11] Let us explore this duality, commonly described as involving some tension between "science" and "art," in a bit more detail.

Medical science is solidly built on the life sciences and has a clear explanatory goal, although it lacks the explanatory power of some of the most developed physical sciences. At least since the discovery of the role of microbes in infectious diseases, it has drawn heavily on findings in biology, which successfully identified entities and processes on cellular, genetic, and molecular levels causally connected to diseases. Even so, medical science is not reducible to the sum of contributing disciplines like biology, biochemistry, and physiology. The aim is to gain a theoretical understanding of biological processes to provide (causal) explanations of disease in exclusively physiological terms. Medical science encompasses clinical research

[11] *Stedman's Medical Dictionary* defines medicine as "the art of preventing or curing disease; the sciences concerned with disease in all its relations," while *Dorland's Illustrated Dictionary* defines it as "the art and science of the diagnosis and treatment of disease and the maintenance of health."

(which investigates aspects of clinical practice, such as the efficacy of a treatment or improvement of diagnosis) and medical laboratory research (which involves research methods and constructing models similar to those of biology and chemistry). The latter aims to enhance those clinical activities, but is not principally focused on the clinical activities (diagnosis, treatment, prevention), which makes it difficult to pinpoint the difference from research in biology or chemistry: parts of medical science are not essentially different from lab science.

Medical practice typically involves personal encounters between medical professionals and patients and aims to improve the health of individuals and groups. It involves prevention, diagnosis, treatment, and promoting health, and it is often described as an "art" in order to stress the "prudential" engagement of physicians that goes beyond understanding and explanation in terms of scientific laws.[12] However, as Aristotle noted, medicine is not pure practice and does not completely correspond to what the Greeks called *technê* (a form of technical skill to produce things), because it also involves command of a body of scientific knowledge of phenomena in terms of causes, effects, and laws.

There are a number of considerations that speak against this way of describing the duality of medicine in terms of a tension between "science" and "art" (see Solomon 2015, 4–9). Rather, medicine might represent, as Gadamer (1996, 39) aptly puts it, "a peculiar unity of theoretical knowledge and practical know-how within the domain of the modern sciences, a unity moreover which as such cannot be understood as the application of science to the field of praxis." But putting the question about tension aside, our opponent could maintain that once this duality is properly taken into consideration, one finds a number of features that lend support for the view that medicine is not a science.

An attempt to show that medicine is not a science could start by acknowledging that medicine *uses* scientific knowledge to support its judgments, but does not qualify as science, because it "is largely observational and functions without the level of certainty essential to science" (Miller and Miller 2014, 151; also Miller 2014). The problem with such an approach is that it sets the bar unjustifiably high in a way that would exclude a large number of fields that are widely recognized as science (e.g., evolutionary biology, climate science, and theoretical physics). Therefore, a better approach for our opponent is to hold on to the starting point that

[12] Later chapters will deal with this question, but it is in its place to briefly indicate how difficult it is to get a grip on what medical practice is.

medicine *uses* scientific knowledge to support its judgments, but argue that the use of scientific knowledge is not sufficient for the label "science."

Consider an example. Imagine that the ambitious coach of a competitive cycle team develops a chemical compound to boost the performance of the athletes on the team. For this, he draws heavily on scientific knowledge about human physiology, in particular about the means by which the number of red blood cells can be augmented in order to increase the transport of oxygen to the muscles. While we may describe his activity as *scientific doping*, his endeavor certainly does not qualify as science. Along similar lines, one might argue that medicine, and clinical medicine in particular, is *scientific*, because it applies knowledge, for instance, from chemistry and biology, but it does not qualify as a science.

The first – and perhaps less decisive – reply could start by granting that in many instances clinical medicine is scientific: it deploys scientific knowledge to support its diagnostic judgments and its therapeutic measures. But, having granted this, a quick comparison with activities traditionally considered as science reveals a flaw in the objection. There are activities in the sciences that we should perhaps refer to as "scientific," because they are characterized by the application of existing knowledge, such as reviewing recent findings in order to design a new experiment. But even if they deserve the label "scientific," these activities are still considered to be genuine parts of science. An activity being scientific, at least in the sense used here, thus does not exclude that activity from being a part of science. But then, even if clinical medicine consisted of merely deploying scientific knowledge (which it does not), that would not be enough to conclude that it is not a part of science. This, however, cannot be said to be true in the doping case, which might be scientific, but is not a part of science.

We might consider the possibility that what is meant by the objection that clinical medicine is scientific but not science, is that clinical medicine is not systematic enough to count as science. In order to show that this is not the case, we may explore the diagnostic process in light of the dimensions of systematicity. Diagnosis starts by systematic description, which involves organizing a range of information provided by the patient and recording the details of symptoms and signs. If necessary, qualitative descriptions are replaced by quantified descriptions, which increase the accuracy of the description. The process is accompanied by an ongoing assessment of whether sufficient information has been collected. On this basis, the next step – the pursuit of an explanation for the condition of the patient – also proceeds in a systematic manner. For example, a differential diagnosis proceeds by hypothesizing a number of underlying conditions as possible explanations of the signs or symptoms, followed by a process of eliminating hypotheses (or at least rendering them less probable) by further tests, experiments, and observation. The aim is to reach the point where the use of an inference to the best explanation allows identifying a condition as the most plausible explanation. This is taken to explain the condition of the patient more adequately than competing hypotheses and allows for a more detailed prognosis.

Contributing to the systematicity of the endeavor, the diagnostic process includes attempts to eliminate or minimize sources of error by using methods that extend from simple checklists and daily joint discussions of cases to using AI algorithms for greater efficiency and accuracy. The use of more complex methods also brings to light high degrees of epistemic connectedness not merely with biology or physiology, but also mathematics and computer science. In addition, regarding the generation of knowledge, clinical medicine (even outside of hospital environments) makes a contribution to broadening the understanding of health and disease. Practicing physicians routinely take part in larger-scale research, for instance, by registering patients into databases, thereby contributing to knowledge production. For instance, in 2014, the Danish Health and Medicines Authority published a document that lists requirements for medical specialists practicing in Denmark. Based on a model originally developed in Canada, the document describes the competencies required of medical specialists in the roles they play (including as medical experts, managers, scholars, collaborators, health advocates, communicators) upon the completion of their training. The document maintains that as a part of their practice "physicians must contribute to the systematic collection, analysis and processing of data with the purpose of launching health-promotion initiatives at institutional and societal level" (Danish Health and Medicines Authority 2014, 28).

Such contributions to systematic data gathering and analysis show how everyday medical practice can be a part of scientific inquiry. Overall, thinking in terms of systematicity, the diagnostic process, at least in its proper forms, qualifies as science in action. But it should be stressed that the criteria of systematicity apply not only diachronically (assessing a practice over a span of time), but also holistically, which leaves room for the possibility that some parts of clinical practice are less systematic than other parts.[13]

[13] Or, for that matter, that some parts of medicine (medical practice) are less systematic than other parts (medical science).

In conclusion, we have provided several reasons for thinking that clinical medicine is not merely scientific, but can qualify as science because it displays systematicity on most dimensions. Of course, our opponent might wonder whether doping could become a science along similar lines. The answer is that there is indeed nothing that would in principle prevent doping from becoming a science. If a number of coaches in similar situations engage in a joint activity that displays systematicity along a significant number of dimensions (descriptions to reduce diversity, epistemic connectedness, striving to attain new knowledge by performing new observations, eliminating sources of error, engaging in predictions of not-yet-observed events, etc.) then it would be justified to speak of "the science of doping."

3.9 Conclusion

Drawing on discussions about the characteristics of scientific inquiry and knowledge, this chapter set out to address a general question about the nature of medicine and sought support for the *Systematicity Thesis*. It is often claimed that science is the preeminent path to knowledge, and as such, we may expect that it displays discernible unifying features. Given that knowledge can be achieved by the exercise of "commonsense" methods, the sciences must possess some special excellence in the acquisition of knowledge. For this, the chapter consulted the literature on the "demarcation" problem in the philosophy of science. It was argued that the failure of well-known approaches should not compel us to abandon the issue, but rather to pose the demarcation question in a different manner. Given the lack of commonalities that characterize the entire range of activities under the conceptual umbrella of science, we should not entertain essentialist expectations and expect to find ahistorical or discipline-independent necessary and sufficient conditions. Instead, science is best seen as a family resemblance concept. The most promising way to consider the sciences as united is not through some intrinsic property, but a *relational property* that only admits differences of degrees to nonscientific undertakings.

In addition to exploring a recent account that highlights systematicity as a necessary criterion for science, our inquiry revealed that medicine meets the requirement for systematicity on all nine dimensions, offering us good reasons to think that it is a science. Of course, the fact that we have not offered both necessary and sufficient criteria limits the strength of this conclusion. However, especially as we were able to show that medicine is systematic on all the dimensions that Hoyningen-Huene (2013) discusses, it seems safe to conclude that it is a good candidate for being a science. At the same time, thinking of medicine in terms of systematicity is helpful. For example, it allows demarcating medicine from homeopathy, which is not systematic, be it synchronically or diachronically. So, while systematicity theory cannot fully establish that medicine is science, it is nonetheless able to demarcate it from at least a prominent pseudoscience, which is one of the reasons why thinking about demarcation is significant in the first place. The main reason why homeopathy does not belong to science is the lack of systematicity on a large number of dimensions. While the reference science (i.e., medicine) exhibits significant increases showing that advancement is possible, homeopathy has produced very few cognitively worthwhile results. Compared to the reference science, homeopathy does not exhibit synchronic and diachronic systematicity, has not attempted to overcome the limitations of everyday reasoning, and remains susceptible to a variety of biases. Surely, the lack of systematicity in homeopathy might not be the *only* reason why it is not a science and there might be other features distinguishing medicine from homeopathy that could contribute to demarcating them. However, as systematicity is a necessary criterion for science, and (in contrast to medicine) homeopathy does not exhibit it, it is sufficient for purposes of demarcation.

While systematicity can be used for purposes of demarcation between science and pseudoscience, it has certain disadvantages. But even so, its focus on relational properties is more promising than those accounts that focus on intrinsic properties and fail due to the immense internal and historical variety of science. In addition to functioning as a *necessary criterion* for science, and being able to exclude fields widely recognized as pseudoscience, it also functions as a criterion for the type of reasoning that produces reliable knowledge. This is crucial, because, lastly, the question "Is X science?" boils down to whether the knowledge X produces is reliable and leads to understanding. With this conclusion about the nature of scientific inquiry and medicine, we may now confront the question of what such inquiry is aimed at and how medicine fits into the picture.

CHAPTER 4

Inquiry in Medical Science The Understanding Thesis

4.1 Introduction

As argued in the first chapters, different strands of criticism converge on fundamental questions about the nature of medicine, on the one hand, and the aim of medicine on the other. The defense of the *Systematicity Thesis* in Chapter 3 helped to clarify the nature of medicine in terms of systematic, scientific inquiry. On this view, scientific inquiries in medicine are goal-directed activities that are continuous with everyday practices of seeking information (see, e.g., Kelp 2021), but they pursue and offer information in a more systematic and fine-grained manner.

While systematicity helps describe the nature of scientific inquiries in medicine, what is their aim? This chapter seeks to make a critical step toward answering this question. Its focus is on the *epistemic* aim of inquiry in medical science, which, as described in Chapter 3, encompasses clinical as well as medical laboratory research, and only counts as properly *medical* (and not biological or something else) if it displays a *practical orientation,* that is, if it is ultimately motivated by contributing to the maintenance of health and the diagnosis, prevention, and treatment of disease.[1]

The main thesis that this chapter puts forward, the *Understanding Thesis,* holds that inquiry aims at *understanding,* while the question of what special kind of understanding is at stake in medicine is the topic of subsequent chapters. Beyond constituting an important step in the overall structure of the book, the *Understanding Thesis* and the arguments presented in its favor have a direct bearing on a key issue in the philosophy of medicine. Some influential figures in this field have argued that due to such practical orientation, inquiry in medicine differs *in kind* from scientific inquiries, leading them to the conclusion that "medicine is not, and

[1] The focus on at least potential benefits for health characterizes not only clinical studies and meta-studies, but also areas of specialization, such as anatomy, physiology, pathology, microbiology, or genetics.

cannot be, a science" (Munson 1981, 189; Pellegrino 1998; Miller and Miller 2014). It will be argued that the practical orientation of inquiry in medicine does not render it different *in kind* from scientific inquiries and does not prevent it from being a science. However, there are important differences in degree, which make a difference for what counts as progress.

4.2 Aim and Constitutive Aim

Medical research is a highly complex activity, and given the range of different types of projects that medical scientists are involved in, one might wonder how plausible it is to assume that it has a single aim. But while it might be true that complex activities and institutions tend to have a plurality of aims, it is nevertheless possible for them to have a single or at least very few *constitutive aims*. For instance, the stated constitutive aim of *NATO* is to guarantee the security of its member states, but what NATO mostly does (e.g., food distribution, nation building, participating in parades, building infrastructure) only indirectly furthers the constitutive aim. An important difference is that while NATO could stop pursuing these nonconstitutive aims without a loss of identity, failing to take steps to protect a member nation from an armed attack from outside the coalition would constitute a violation of its constitutive aim and lead to a loss of identity.

The investigation here is focused on the constitutive aim of scientific inquiry in medicine, which also governs what counts as progress. So, if X is the aim of inquiry, then medicine makes progress when X accumulates or increases.[2] For something to be the aim of a progressive activity, two conditions have to be fulfilled. First, the aim has to be something that can actually be achieved within the boundaries of the activity, at least if performed according to its rules (Rowbottom 2014).[3] This means that if a certain goal (e.g., eternal life) is unachievable, then it cannot figure as the aim of inquiry in medicine. While it is difficult to determine how high the minimum probability of achieving the aim has to be when properly engaged in the activity, we may say that if X is the aim of Y, then proper engagement in Y must offer some probability of achieving X.

[2] For a helpful general discussion of this matter, see Bird (2007).

[3] In general terms, the aim of inquiry cannot be to offer a "theory of everything" (formulated as a finite number of principles) if such a final theory cannot be given or if it is logically impossible for a single theory to describe all phenomena in the universe. In the same way, the aim cannot be to offer a complete chart of "nature's joints" if there are no such "natural" divisions to identify. Pursuing an aim that is known to be unattainable within the framework of an activity would be irrational.

Second, the aim must fulfill success criteria intrinsically linked to the aim of the activity, because the aim of an activity sets conditions for the success of that activity. There are two subconditions here. First, an activity is progressive if individuals or groups properly engaging in the activity regularly achieve the aim or at least make reasonable steps toward it. An activity is not progressive if there is no approximating movement toward the aim for extended periods of time. Second, whether the aim is achieved or approximated, it has to occur in an appropriate manner (i.e., through the application of the proper rules and methods) to count as progress. If it occurs by chance, manipulation, or fraud, then it would be unfitting to speak of true success or progress.[4] To take an example from the history of medicine, in the early nineteenth century Johann Friedrich Meckel developed the theory that human embryos pass through successive developmental stages that correspond to adult forms of less complex organisms in the course of development (fish, amphibians, reptiles, etc.). The theory predicted that at a certain stage of their development, human embryos would have gill slits. The later discovery of embryo neck slits that resemble gills persuaded some to accept Meckel's theory. The hypothesis was generated by relying on false background assumptions, and the process that led to accepting the theory did not meet the standards of systematicity. Although Meckel's theory made a risky and correct prediction, it does not count as progress.

To sum up, it is possible for a complex activity like medical research to have a single constitutive aim, which determines not only what we may legitimately expect medicine to achieve, but also what counts as progress. Medical research must offer some probability of achieving its aim and requires that individuals or groups properly engaging in medical research at least make reasonable steps toward the aim.

4.3 Truth, Knowledge, and Understanding

So what is the epistemic aim of scientific inquiry in medicine? According to a plausible suggestion, the aim is simply to discover truths about health and disease and correct past errors (e.g., false beliefs about diseases like scurvy or depression being caused by humoral imbalance) that were based on tradition, cognitive mistakes, ideologies, or religious dogma.

[4] Just as winning the Tour de France while under heavy doping does not count as success, offering a correct answer or prediction in science while not following its epistemic guidelines is not sufficient for progress.

Correspondingly, progress consists in a cumulative acquisition of true beliefs. At first glance, this is a reasonable proposal. After all, it is often said that scientific inquiries are in the "truth business" (Lipton 2004; Pennock 2019, 26),[5] and it is difficult to imagine that contemporary medical science would be able to achieve what it does if its claims did not at least roughly correspond to how the world actually is.

Nonetheless, the acquisition of true beliefs does not seem sufficient to constitute progress.[6] For example, consider a situation in which scientists come to believe a theory by using unreliable methods and, by chance, it turns out that the theory is actually true. Take the discovery that the bacterium *Helicobacter pylori* plays a causal role in gastritis and peptic ulcers. It clearly counts as true progress, and for their discovery of the role of this bacterium, Barry Marshall and Robin Warren won the 2005 Nobel Prize in medicine. Marshall and Warren noticed an association between inflammation severity and the concentration of bacteria in biopsy specimens. They initially failed to culture bacteria from biopsies, until, due to sheer luck, the incubation period in one set of plates was tripled, permitting optimal growth (Kyle, Steensma, and Shampo 2016). This is of course not the full story,[7] but imagine if Marshall and Warren had reached the same conclusion based on observing a set of plates that were by some fluke accidentally contaminated with *Helicobacter pylori.* In that case, the scientists would have acquired a true belief, but it would not have counted as genuine progress. What would be lacking is suitable justification for holding the relevant belief. In other words, the belief that Marshall and Warren would have acquired in our thought experiment would not qualify as *knowledge.*

What we learn from these considerations is that progress not only requires that our beliefs and theories be true but that we have attained adequate reasons for forming them. If this is correct, then it seems safe to conclude that the aim of inquiry is not merely truth, but *knowledge* (achieved by reliable means, for instance, by fulfilling the requirements of *systematic* inquiry), which would mean that progress consists in the increase not of true beliefs, but of knowledge. This conclusion is very

[5] Pennock (2019, 6) notes that when scientists maintain that scientists seek truths about the world, they are using a "humbler sort of empirical truth."

[6] See Bird (2019a, 175–6) for a helpful discussion of an example from physics.

[7] To show that *H. pylori* caused the inflammation (and to establish it is not the inflammation that allowed *H. pylori* to grow), Marshall swallowed a culture of the bacteria and developed gastritis. His subsequent stomach biopsy showed the presence of *H. pylori.* Subsequent treatment with antibiotics led to prompt recovery, fulfilling Koch's postulates (Kyle et al. 2016).

much in accordance with what we have said about systematicity in Chapter 3: systematicity protects against various kinds of biases and allows reasoning processes that reliably lead to the acquisition of knowledge.

So perhaps progress in medical science is constituted by the accumulation of knowledge with the ultimate aim to attain a complete "map" of knowledge of all facts in the realm of medicine, which could serve all kinds of purposes that require information. But such a view would ignore a number of issues. First, what is possible to know about these matters, even if they only constitute a tiny part of nature, is too vast ever to be completely assembled. Even if we came close, such an ideal map would be far too detailed to function as a good map. What makes a good map ultimately depends on what we are trying to achieve when using it: it is only good if it successfully picks out *significant* aspects determined by our purposes.[8] Second, it would overlook that there are vast amounts of insignificant knowledge that we could accumulate, such as knowledge about the amount of hemoglobin in the left side of the *orbicularis oculi* muscle tissue in my face on the first day of the spring semester.[9] Dedicating valuable research time to such efforts would be hard to justify, and success in such inquiries would not count as genuine progress.

If this is correct, then the aim of inquiry in medicine cannot be the mere accumulation of knowledge, but rather a selective acquisition that privileges certain types of *significant* knowledge. In light of the near infinite number of possible questions that inquiries in medical science might investigate and the near infinite number of bits of knowledge that can be attained, deciding what significant knowledge is worth seeking is an unavoidable part of the scientific process (Dupré 2016). Here, it is worth noting that what counts as significant knowledge cannot simply be "read off" nature. One might think that medical science is ordered in a hierarchy such that regularities on a higher order of the hierarchy can be derived from the laws on the lower orders, and that significant knowledge is

[8] For example, the topographical map of San Francisco contains all the information about changes in elevation in the city, but no relevant information that will help you get to the Twin Peaks Tavern. Maps have to be selective to be able to efficiently serve goals at all, while not representing, simplifying, or offering a distorted representation of features that are not of interest. A similar point holds in the case of science. It does not merely aim at gathering knowledge (which would be impossible and uninteresting), but *significant* knowledge, the significance of which is relative to specific interests: "significant science must be understood in the context of a particular group with particular practical interests and a particular history" (Kitcher 2001, 61).

[9] Similarly, an inquiry aiming to offer relatively reliable predictions of the amount of hair on men in their seventies might only be marginally more important than offering empirical evidence that a serious knee injury is a disadvantage for jogging.

therefore located nearer to the bottom of the hierarchy. In other words, significant knowledge is knowledge of laws or natural kinds studied in medical molecular biology and physics that grasps the structure of nature on a basic level. However, such fields of inquiry focusing on the molecular level do not operate with general laws that apply to the entire range of phenomena of interest to medicine, and thinking in terms of natural kinds requires making assumptions about the structure of the world that cannot be readily "read off."[10] Instead, as we shall see in more detail, what counts as significant knowledge is indissolubly tied to practical matters.

4.4 Understanding as the Aim of Inquiry

There is perhaps a better candidate than knowledge for *the kind of fundamentally valuable epistemic state* that scientific inquiry in medicine aims at. In recent debates (see, e.g., Grimm 2017; 2021), some philosophers have argued that, compared to knowledge, understanding is characterized by "a more intimate epistemic acquaintance" (Strevens 2017).[11] While understanding can refer to (a) the phenomenology of understanding (i.e., the feeling or sense of understanding that may accompany an explanation), (b) understanding a theory (i.e., the ability to use it), or (c) understanding a phenomenon by having an adequate explanation of the phenomenon (De Regt 2017, 24), we will be dealing with understanding in the sense of (c), which often produces a sense of understanding (a) and involves (b), but need not do so.

Chapter 5 will offer a detailed discussion of understanding; for our purposes in this chapter, a brief sketch of three important aspects of understanding is sufficient. First, we can know *p* (or why *p*) without understanding *p* (or why *p*). Achieving understanding, and an additional component consisting in "grasping" dependency relations, enables one "to

[10] In more general terms, along with the demise of the idea of a unified science (see, e.g., Cartwright 1999; Dupré 2004), it has become doubtful that nature offers us clear guidelines of what is significant.

[11] In epistemology, the focus is shifting away from analyzing knowledge as justified true belief and evaluating Gettier-type counterexamples, and, among other developments, the notion of understanding is being rehabilitated (Grimm 2017; 2021). In philosophy of science, the emphasis has long been on "explanation," and it was assumed that understanding is little more than the philosophically uninteresting psychological side effect of explaining. Evaluating explanations based on whether they yield understanding would introduce major insecurities: due to biases and the lack of cognitive sophistication, poor explanations might produce a sense of understanding while good explanations may fail to. Worse, the sense of understanding ("Eureka!") risks misleading scientists into supposing that the correct explanation has been discovered (Trout 2002).

anticipate how changes in one part of the system will lead (or fail to lead) to changes in another part" (Grimm 2011, 89; 2014; Strevens 2013). Second, understanding is inseparable from explanation. According to a standard account "S understands why *p* if and only if there exists some *q* such that S knows that *q* explains why *p*" (Khalifa 2017, 18).[12] Understanding why *p* involves being able to explain why *p* occurs instead of something else, which requires knowledge of causal dependencies relevant to *p*.[13] Third, understanding is *factive*, but not to the same degree as knowledge. In many cases, understanding *p* (i.e., objectual understanding of a subject matter) or understanding why *p* does not require that all propositions constituting one's representation of the subject matter or the explanation be true. Understanding admits of degrees, and as long as the falsehoods are tangential, one's understanding is degraded but not lacking.

For a brief illustration of the difference between causal knowledge and understanding, Duncan Pritchard (2010) considers a situation in which a house burns down while a family is away. Upon returning to the scene, the curious young boy of the family asks why it burned down, and the fire chief tells him that it was because of faulty electrical wiring. By accepting the testimony from a reliable source, it seems that the boy has gained causal knowledge, but it is not enough for understanding without some idea of *how* the cause (the faulty wiring) might bring about the effect (the fire). Considering similar differences, some have argued that the aim of inquiry is often understanding, not knowledge (Pritchard 2016; Kelp 2021). The knowledge that the boy attained will not properly close his inquiry, and he will probably carry on seeking information. His inquiry will reach its natural end when he comprehends not just *what* the relevant cause is, but *how* cause and effect are related. Until then, he will lack the ability to answer what-if-things-had-been-different questions and to anticipate the effects of certain conceivable interventions (Grimm 2011).

Pritchard's case helpfully contrasts knowledge and understanding, but one might doubt its relevance for establishing the aim of inquiry in medical science, in part because it uses an example of knowledge

[12] There are nonexplanatory forms of understanding, such as those expressed in sentences like "I understand what I need to do." Moreover, understanding can be nonexplanatory – a person may understand how a device works by way of her ability to correctly manipulate that device while lacking explanatory resources (Kvanvig 2009; Lipton 2009).

[13] Dellsén (2016, 74) argues that correct predictions can express understanding even if any explanation is lacking. However, if one grants the possibility of partial understanding, which Dellsén does, it is not clear why we cannot speak about partial explanation in a case with partial understanding. This would be in line with the general emphasis in the literature on the inextricable relation between understanding and explanation.

acquired by testimony. This is, of course, profoundly dissimilar to knowledge acquisition in scientific inquiry, and for this reason we need to supplement our considerations by a brief discussion of an example from medical science. A good example of an utterly false theory leading to correct results is Charles Locock's mid-nineteenth century discovery of the anticonvulsant effect of potassium bromide. Locock, a physician working in London, shared the widely accepted theory among his contemporaries of a causal relationship between masturbation, convulsions, and epilepsy (Ban 2022). As bromides were known to reduce the sex drive, Locock reasoned that the ingestion of potassium bromide would control convulsions by reducing the rate of masturbation. His account of the drug's effectiveness was published in *The Lancet* in 1857, and subsequent independent studies confirmed potassium bromide's antiepileptic efficacy, albeit evidently not by reducing masturbation frequency. Through observations and inference to the best explanation, Locock had attained knowledge that potassium bromide reduced convulsions, and such knowledge allowed the introduction of a relatively effective antiepileptic treatment into medical practice.

Still, in an important sense, such causal knowledge does not properly close the inquiry, which would require grasping a correct explanation and attaining *understanding* of what happens and *how* cause and effect are related. Locock did not understand why potassium bromide was effective, why it failed to be effective in some people, and so on. This meant that he lacked the ability to improve the efficiency of the intervention, since he was unable to counterfactually anticipate the effects of changes he could have made with respect to the treatment. More precisely, the lack of understanding means that Locock was unable (i) to predict the changes that would occur if the factors cited as explanatory were different and (ii) to correct conclusions about similar cases in slightly different circumstances by engaging in counterfactual reasoning.

4.5 Progress: Knowledge or Understanding?

One might accept that understanding is one possible aim of inquiry, but why should one accept the *Understanding Thesis*, claiming that understanding (and not knowledge) constitutes the aim of inquiry in medical science? While keeping in mind that all understanding is also knowledge, the *Understanding Thesis* has a crucial advantage compared to construing progress in terms of knowledge accumulation. The advantage with comprehending progress in terms of increased understanding is that it is resistant to difficulties that accounts of scientific progress measured in terms of knowledge accumulation run into (see, e.g., De Regt and Dieks 2005; Elgin 2007; 2017; Potochnik 2017).[14] First, traditional accounts have problems explaining the significance of certain pragmatic virtues (e.g., simplicity) that do not affect the truth of claims, theories, and explanations. In contrast, an account of progress based on the notion of understanding does not face this problem, as these pragmatic virtues clearly affect the ability to understand (Dellsén 2016).[15] Second, traditional accounts of progress as knowledge accumulation have problems explaining abstractions, approximations, and idealizations. For example, in medicine, physiological accounts often offer idealized and simplified descriptions of organs and their functions (Ereshefsky 2009). These provide computational tractability and improve understanding, but they also include aspects that are, strictly taken, inaccurate or false. However, such falsehoods are, as Catherine Elgin (2017) puts it, "felicitous": although they involve false representations, they also exemplify significant aspects of phenomena in an epistemically tractable manner. A number of philosophers have argued that science can increase understanding and contribute to progress even if it involves departing from the truth (Elgin 2009a; 2009b; Potochnik 2015; Strevens 2017).

On an account of progress in terms of knowledge, the presence of manifest falsehoods seems incompatible with progress. However, an account of progress in terms of understanding fares better here, since understanding is compatible with a limited number of falsehoods, which are outweighed by practical advantages.[16] The latter is not susceptible to such worries, because, in contradistinction to knowledge, understanding is only quasi-factive: it can survive false beliefs if they are not absolutely vital to the phenomenon in question. For an example, we may return to Ignaz

[14] Limitation of space prevents considering other accounts, such as the idea that progress in science occurs when theories come nearer to the truth (verisimilitude) (Popper 1979) or when it accumulates solutions to scientific puzzles that are neutral about questions of truth and knowledge (Kuhn 1962; Laudan 1984). For a critical review, see Bird (2007).

[15] Even if one holds onto the view that science is crucially concerned with explanations, the conceptual link between explanation and understanding requires a full account of explanation to say something about understanding.

[16] Strevens argues that idealized models can provide understanding, but in a somewhat more limited way, showing why some causal factors are difference-makers and others are not (Strevens 2017). Angela Potochnik (2015; 2017, 102) holds that while idealizations cannot be true or approximately true, they can be epistemically acceptable. Because such idealizations are rampant in science and they always detract from the truth, truth does not seem the appropriate aim of science. However, given that idealizations can support understanding, it is more adequate to suppose that understanding is what science aims at.

Semmelweis's examination of childbed fever in two maternity wards. Semmelweis noticed that medical students examined patients in the high mortality ward, and after considering other hypotheses, he came to the conclusion that childbed fever was caused by "cadaveric matter" that the medical students were infecting patients with. Strictly taken, this is false: childbed fever was not caused by "cadaver material" (there is no such thing), but bacteria, which Semmelweis had no knowledge of. Nevertheless, it is hard to deny that some progress occurred and an increase in the (objectual) understanding of childbed fever had been obtained.

In all, as opposed to knowledge or truth, the epistemic aim of inquiry in medical science is best comprehended as understanding. Comprehending progress in terms of increased understanding has the additional advantage of being resistant to some of the problems that haunt accounts of scientific progress measured in terms of knowledge accumulation. Pursuing understanding has a profoundly *practical orientation*, seeking to contribute to maintenance of health and the diagnosis, prevention, and treatment of disease. When we say that it is with such an aim that inquiries such as clinical trials attempt to understand relationships of causal dependence with respect to diseases, there is thus an intimate link between our ability to understand and our *capacity as agents* to fight disease, to maintain and improve health, and to control the environment if relevant for matters of health. It is not difficult to see that even if no cures are available for particular diseases like cancer or arthritis, the increased understanding of their nature that we have obtained renders them more manageable than they were to our ancestors, thereby increasing our agentive capacities as well as our ability to determine our own values and ends (i.e., self-determination), which we standardly take to be some kind of minimal condition for autonomy.[17] On a more general level, self-determination requires some degree of comprehension of causal dependencies, as we need to be able to establish whether our acts can succeed in realizing the values and ends we pursue. With a lack of such understanding and in the presence of false beliefs (e.g., it is possible to cure these diseases through bloodletting or purging) agential capacity for self-direction would be compromised: we would not know why our actions are ineffective and would remain unable to alter our behavior to better achieve our ends.

[17] It bears emphasizing that this is a minimal condition, as autonomy involves much more, including the possibility to participate in and contribute to collective undertakings.

4.6 A Difference in Kind?

We may now return to a critical issue that some influential voices in philosophy of medicine have raised. Munson (1981), Pellegrino (1998), and Miller and Miller (2014) have argued that due to its *practical orientation*, inquiry in medicine differs *in kind* from scientific inquiries. More precisely, these authors argue that because inquiry in medicine (a) is driven by distinct aims, (b) has different criteria for progress, and (c) has different moral commitments, the conclusion must be that "medicine is not, and cannot be, a science" (Munson 1981, 189; Pellegrino 1998; Miller and Miller 2014). Let us first review these points in more detail.

As to (a), these authors claim that the aim of science is "pure" in the sense that it consists in "the acquisition of knowledge and understanding of the world" (Munson 1981, 190), while the final aim of inquiry in medical science is fundamentally *practical*. With respect to (b), proponents of this view accept that medical research projects might in some cases look identical to nonmedical ones, but scientific progress occurs when research projects increase understanding in their respective fields, while progress in medicine requires increasing understanding that holds the promise of something close to assisting the promotion of health. Thus, progress in science is measured in "pure" epistemic terms, while progress in medicine is measured in practical terms. Finally, with respect to (c), the difference in aims is also reflected in different responsibilities and moral commitments. "Pure" science is not accountable to moral, social, and political standards beyond communicating findings in a sincere fashion and pursuing inquiries in a way that does not harm participants. In contrast, in medicine, there is a moral commitment to promote the health of individuals and populations.

In the following sections, the chapter will offer a discussion of these three points, arguing that the "purist" picture that Munson and Pellegrino propose is flawed and fails to show that scientific inquiry in medicine is somehow different in kind due to its practical orientation. Importantly, contributing to a fuller characterization of the defining features of inquiry in medical science, the discussion will help highlight significant differences in degree.

4.6.1 Distinct Aims?

In contrast to what the "purist" picture maintains, practical orientation and interest in intervention and manipulation is not only not restricted to medicine, but is also built into the very notion of explanation in a wide range of scientific contexts. As James Woodward (2003, 11) maintains, "had we been unable to manipulate nature – then we would never have developed the notions of causation and explanation and the practices associated with them that we presently possess." Moreover, espousing the view that causal relationships holding between variables are a matter of counterfactual dependence, Woodward (2003, 13; 2015) stresses that explanations in science "provide understanding by exhibiting a pattern of counterfactual dependence of the special sort associated with relationships that are potentially exploitable for purposes of manipulation and control." Ultimately, the epistemic significance of discovering causal dependences stems from the *practical* significance of such understanding for manipulation, control, and prediction. As Walker and Gopnik (2013, 342; Gopnik 1998; 2000) put it, causal understanding is "useful, both ontogenetically and evolutionarily, because it allows for a special kind of counterfactual called an intervention. Once you know how one thing is causally connected to another, this knowledge allows you to deliberately do things that will change the world in a particular way."

Consider, for example, the important task in scientific inquiries of distinguishing causal and correlational relationships. To explain one reason why the difference matters, consider that while X's being the cause of Y implies that manipulating X is a way of manipulating Y, X's being merely correlated with Y fails to imply that manipulating X is a way of manipulating Y.[18] To take an example, there is a correlation between chocolate consumption in a country and the cognitive performance of its population, measured by the total number of Nobel laureates per capita (Messerli 2012). Presumably, the correlation will not be invariant under attempts to use the former to manipulate the latter, and the correlation probably arises due to some common cause (e.g., the total wealth of the country) that affects both chocolate consumption and the number of Nobel laureates. The point is that in contrast to causal relationships of dependence, correlational relationships are not exploitable for purposes of prediction and intervention.

[18] Manipulation or intervention in this sense does not require reference to human action. In Woodward's manipulability account of causation, an intervention is an idealized experimental manipulation deployed to determine whether a variable is causally connected to another variable (Woodward 2015, 311). More precisely, "X causes Y if and only if there are background circumstances B such that if some (single) intervention that changes the value of X (and no other variable) were to occur in B, then Y or the probability distribution of Y would change" (Woodward 2010, 290).

The "map" of relationships of dependence that scientific inquiries offers us is – unlike everyday practices of inquiry – *systematic* and *fine-grained* and differentiates (a) contributing and direct causes, (b) specificity and stability, and (c) causal vs. constitutive relationships of dependence. Let us see what they amount to while keeping in mind how, ultimately, these distinctions make sense from the vantage point of *practical* significance, enabling more fine-grained manipulation, prediction, and control.

(i) *Contributing* and *direct* causes offer different opportunities for manipulation. X is the direct cause of Y just in case there is an intervention on X with respect to Y, which will alter Y (or its probability distribution) while all other variables are kept fixed. X is the contributing cause of Y just in case there is a directed path from X to Y (with a direct causal relationship between each link) and there is a possible intervention on X with respect to Y that alters Y while all other off-path variables are kept fixed (Woodward 2003, 59).

(ii) More *stable* and *specific* causal relationships are likely to offer more opportunities for fine-grained manipulation and control than those exhibiting less stable and specific relationships. The relationship $X \rightarrow Y$ can be more or less *stable*, contingent on the range of background conditions under which it obtains,[19] and more or less *specific*, depending on the grain level of counterfactual dependencies between X and Y. A high specificity means that interventions on X allow a fine-grained intervention on and modulation of Y.[20]

(iii) Scientists often experimentally distinguish between causal dependence and constitutive dependence (which holds between a system's macro-level activity and the behaviors of its components' relations) in terms of manipulability (see Craver 2007; Craver and Darden 2013). If X causes Y, then the relationship of dependence between X and Y is asymmetrical (Y depends on X but not vice versa) and X and Y are mereologically independent entities.

[19] Lewis (1986) discusses this issue in terms of "sensitivity," while Woodward (2010) uses the term "stability."

[20] To illustrate causal relations with high and low specificity, Woodward (2010) considers a radio equipped with a tuning dial and an on/off switch. The switch and the dial are specific to their effects to different degrees. The influence of the switch is relatively nonspecific with respect to whether a station is heard, while the dial is a relatively specific cause of which station is heard. Manipulating the switch will have an all-or-none effect, while manipulating the dial will exert a relatively fine-grained causal influence, taking on many possible values resulting in differences in the station that is heard.

> If X constitutes Y, then X and Y stand in a symmetrical relationship of dependence (Y depends on X and vice versa) and are spatiotemporally overlapping entities. While the asymmetrical nature of causal relations means that it is only possible to intervene unidirectionally to change X and thereby change Y but not vice versa, the symmetrical nature of a constitutive relation enables bidirectional manipulability and mutual difference-making.

Of course, this brief sketch cannot offer a full account of the entire range of relations of dependence that scientific inquiries unearth. However, it is sufficient to highlight how charting correlational, causal, and constitutive relationships enables more comprehensive and effective manipulation, prediction, and control. The "purist" picture that Munson and Pellegrino propose fails to show that scientific inquiry in medicine is somehow different in kind due to its practical orientation. Stretching from everyday contexts to scientific inquiry in medicine, and science in general, our interest in inquiry and understanding is motivated by our practical interest in prediction and control, which can contribute to enhancing our capacity to function in the world as agents.

Having a good understanding of soil fertility would be something different for a farmer and for a biologist, and simply stating that one of them has a better understanding than the other is problematic, in part because it does not take into account the different interests they have in understanding soil fertility. In contrast, it does not seem problematic to claim that the biologist has a better explanation, but that only means that the biologist's understanding includes more and more fine-grained causal maps that allow more fine-grained interventions.[21] The difference in their understanding of soil fertility is a product of being motivated by different interests. Similarly, what it means to adequately understand HIV will be different for a physician, a psychologist, and a social worker, depending on how it is related to their ability to intervene in order to ameliorate the condition of the HIV-infected patient.

Of course, in contrast to questions in medical science such as "How do cancers develop?" it may be difficult to see the practical interest in seemingly "pure" theoretical questions, for example, in biology, such as "How do organisms develop?" Nonetheless, to strengthen our point, we may invoke Kitcher's (2001, 78–82) "significance graphs," which depict

[21] However, this is not the case if we compare the leading biologist to her first-year student. Here, we can say that her understanding of X is *better*: it involves, among other things, grasping many more causal factors than is possible for her student.

how different scientific activities inherit their significance. The graphs trace connections between research projects, claims, questions, and problems within a scientific field. The main point is that once such connections are properly traced, it becomes clear that significance does not simply flow from seemingly "pure" theoretical questions, but also from practical matters, such as questions related to the potential improvement of livestock and the use of animals as disease models or as drug factories. This casts doubt as to whether a scientific endeavor can be "pure" to a degree that would mark a difference in kind to "applied" endeavors.[22]

Munson and Pellegrino might object that unlike medicine, scientific inquiries also aim to understand and explain phenomena for which control and manipulation are not relevant or no longer possible, such as in the case of causal explanations of past events like the extinction of the dinosaurs. But this does not pose a problem for the account offered here. First, we may highlight that research in medicine aiming to understand traits associated with vulnerability to disease combines proximate explanations with ultimate (evolutionary) explanations to provide a full picture. Second, the interventions need only to be conceivable (not feasible): evaluating whether X and Y stand in a causal relationship only depends on what *would* happen to Y *if* the intervention were performed, not whether an intervention is actually performed. Quite simply, playing out "what-if-things-had-been-different" scenarios represents an extension of our interest in manipulation and control from cases in which manipulation is possible to cases in which it is not. Third, we could accept the objection and the conclusion that scientific inquiries are broader than those in medicine, interested also in understanding and explaining matters that could not possibly be of practical interest. However, without further argument, it is hard to see how accepting that science has broader aims would force us to accept that medicine displays a difference in kind.

Obviously, increases in understanding do not always directly increase our agency. In many cases, we gain understanding of key aspects of a disease without being able to intervene on the disease based on such understanding, for instance, due to a lack of technological advances. For

[22] Here, Munson and Pellegrino might object that Kitcher's significance graphs track the ultimate source of significance to broad questions that emerge from "natural curiosity," which seems to allow for holding onto some idea of "pure" epistemic significance. However, satisfying curiosity is itself not an epistemic end, but a *practical* end. If what makes something epistemically significant is determined by the degree to which it satisfies curiosity, then epistemic significance hinges on practical significance. So we may agree that curiosity bestows significance; curiosity is deeply connected with our practical interests and activities as agents.

example, this was the case at the beginning of the twenty-first century with important advances in our understanding of the mutational profiles and other alterations of cancer genomes (see, e.g., Mardis and Wilson 2009). Even if a better understanding of the roles of genomic variation and tumor mutational burden with respect to disease progression did not allow for direct intervention on the "micro-level" underlying biological processes, it proved to hold diagnostic and predictive value, thus enabling supporting interventions on the "macro-level" that indirectly contribute to increasing agency. Moreover, also indirectly contributing to agency, increased understanding can correct false beliefs about the disease (e.g., its nature, etiology) that treatment options, diagnosis, and prognosis are based on.

4.6.2 Different Criteria for Progress?

Pellegrino and Munson argue that there is another way in which the practical orientation of inquiry in medicine renders it different *in kind* from scientific inquiries. While increases in understanding qualify as progress in scientific inquiries, this is not the case in medicine. To see why, consider the case of the CIA-funded project MK-ULTRA, which during the 1950s and 1960s searched for mind control drugs or techniques that could be weaponized. Now considered to be a significant example of human rights abuse, it involved using psychedelic drugs, isolation techniques, and radiological implants on unwitting subjects. MK-ULTRA consisted of more than 150 different secret projects that were contracted out to different universities and research organizations, many of which had no awareness of the nature of the project (Kinzer 2019, 478 and 270). Many of the subprojects were pursued in medical university settings by medical doctors, who focused on the chemical code of LSD or how hypnosis and drugs might work in combination (Kinzer 2019, 182).[23]

While the research performed in the relevant medical settings has surely increased understanding, maintaining that there is progress *in* medicine in such cases neither sits well with intuitive verdicts nor with the commonly accepted view of the practical orientation of medical research, as discussed earlier. Even if the discoveries might turn out to be medically relevant, such cases do not count as making progress *in* medicine. There is thus an important constraint on progress that Munson and Pellegrino might argue renders medical research different in kind.

[23] Lastly, the project was unsuccessful, in that the materials and techniques proved far too unpredictable in their targeted effect (Kinzer 2019, 406).

In reply, we may acknowledge a key difference, but argue that there is a sense in which even outside of medicine, it is not clear that such cases amount to progress – or at least progress in its fullest sense – even if they involve a growth in understanding. Of course, such intuitions may not be widely shared. But there are reasons to be skeptical about progress here, because such research projects are *self-defeating* with respect to both the aim and the nature of science. First, to the extent that the inquiry will produce understanding that will foreseeably be used to render people apathetic, docile, and uncritically trusting of authorities, pursuing the inquiry would be self-defeating in the sense of being detrimental to the normative source that affords scientific inquiries their significance. An endeavor that erodes the very source that renders it significant is self-defeating, and it is therefore hard to see how it could be progressive. Second, in the context of our defense of the *Systematicity Thesis*, we have seen that the reliability of scientific knowledge essentially depends on conducting the inquiry in a manner that respects the standards of systematicity. However, systematicity requires a minimum of freedoms for researchers, so inquiries that will foreseeably be used to promote forms of society that do not allow systematic inquiry (e.g., anti-democratic, authoritarian societies) would be self-defeating.

Thus, intuitive verdicts and independent reasons speak against maintaining that there is genuine scientific progress in cases like MK-ULTRA. However, two potential objections arise. First, one could stress that there is a conflict: while we claimed that if X is the aim of science, then science must be making progress when X increases, we now imply that not every increase in understanding will be sufficient for progress. A simple reply is that progress is long-term, whereas an increase in understanding is short-term. This would be consistent with the fact that it would be strange to call every single and insignificant step in an inquiry progress, which is what we would be forced to do if we assumed a narrow *progress-from* (a certain initial state of understanding) model.[24] Instead, the idea is that the MK-ULTRA example might count as a short-term increase in understanding, but it is unsustainable, because in the long run it will be self-defeating.

[24] Progress always involves some kind of movement from a point of departure D, and it can be understood in two ways. If the aim is to get from D to a particular place P, which is fixed, then progress is measured in terms of increased proximity to P. If the aim is to advance as far away from D as possible, then progress is measured in terms of the distance from D to one's current location. Scientific progress might be understood along both lines, but while the second option risks being too imprecise (surely, not all departures from a starting point will count as progress), the first option risks thinking of science as moving toward some "final state" (i.e., "Nature's Book" containing all the true sentences about the world) (Kitcher 2015; 2016, 202).

At this point, Munson and Pellegrino might object that the view proposed here mistakenly collapses scientific progress into societal progress. However, the claim here is less demanding. It is merely that progress is not consistent with self-defeating research deliberately contributing to a decrease in the agential conditions of autonomy. This means that a great number of scientific inquiries will be able to pass the test of progressivity, even if they impact society in ways that would not support or would even be detrimental to the interests of many. Such a view is not as ambitious as Kitcher's account of "well-ordered science," in which scientists engage in "ideal deliberation" under mutual engagement to predict which inquiries would best serve the interest of the community. Well-ordered science serves society's fully considered interests and likely excludes choices that benefit only a small segment of affluent people or subgroup of scientists, while neglecting the interests of the majority.[25]

The view of progress proposed here is in line with the work of those who stress that science is fundamentally embedded in society such that its internal norms are not detached from society's normative landscape. It implies that progress is not neutral on moral aspects, and this is consistent with the role and responsibilities we usually assign to large-scale social and institutional enterprises financed by democratic states. Given that a main objective of democratic states is to promote the autonomy of citizens, publicly funded scientific enterprises must align with expanding the range of opportunities open to its citizens, enabling them to govern (see Anderson 1995, 148–9).[26]

4.6.3 *Different Commitments?*

Perhaps Munson's and Pellegrino's "purist" view is right in claiming that inquiry in science and inquiry in medicine involve different responsibilities

[25] All representatives present their initial investigative preferences, inform others about the epistemic and practical significance of these preferences, and discuss until the conferees are able to reach a decision on the inquiries that they, as a community, are interested in pursuing. This deliberative process is ideal, which means that it is not necessary to actually conduct it (and indeed it would be impossible to), but it correctly predicts which inquiries serve the community's interests. A science is well ordered if it pursues those questions that ideal deliberation under mutual engagement would have identified (Kitcher 2011, 114–15). Kitcher admits that an actual conversation of this type is not possible – nor are the cognitive tasks of including the entire world's population or representatives of the interests of future generations. Nevertheless, the unattainable ideal serves as a guide to achieve ever better approximations (Kitcher 2001, 125).

[26] This does not speak for a tighter political grip on science, or for eliminating peer review in determining the allocation of funding, which would run the risk of devoting inquiry solely to answering questions of immediate practical concern. This would risk rendering it short-sighted, excluding, perhaps, more theoretical lines of research with long-term practical benefits.

and moral commitments. After all, inquiries in science do not share the commitment to promoting the health of individuals and populations that characterizes medical science. So, unlike science, medical science is a "moral enterprise" at least in some weak sense. However, the main point in what follows is that the dissimilarities do not warrant inference to a difference in kind.

For this, let us start with some considerations that the "purist" view would probably agree with. Heather Douglas argues that just like agents in everyday life, scientists can be held responsible for both the intended and unintended consequences of their actions, and they may be blamed for being negligent (not considering possible consequences) or reckless (disregarding potential harms) (see Douglas 2003). Moreover, scientist have a responsibility to refrain from harming participants, honestly report observations and experimental results, and, with respect to the methods of inquiry, uphold the standards of systematicity in their research.

While the "purist" view claims that science is not accountable to other moral, social, and political standards, the view proposed here suggests a more complicated picture. On this view, for both medical scientists and scientists working in nonmedical fields, the requirements of systematicity lead to other responsibilities such as responding to valid criticism and addressing evidence that runs contrary to one's views.[27] Moreover, they do not merely bear responsibility for upholding the standards of systematicity once the research project has started, but also for applying the same type of systematic reflection to *the choice of inquiry* and research goals. It is natural to think that such reflection is a proper part of scientific activity and that the standards of systematicity also apply to it, because agenda-setting considerations help ensure the significance of the research goals and the anticipated increase in understanding. After all, the kind of reflection involved in choosing an appropriate research trajectory (including considering existing knowledge, possible measurement procedures, feasibility, prospects for the success of competing strategies, and counterfactual scenarios) would qualify as a proper part of science if it occurred during an inquiry.

In addition, as a consequence of true progress not being consistent with certain research trajectories that deliberately contribute to decreasing the conditions of agency and autonomy, scientists have an obligation to systematically assess not only the expected epistemic gain in understanding

[27] Although it is not clear that this directly follows, the responsibilities may also include serving as a peer reviewer.

but also what such an increased understanding would mean to society at large. The requirements of systematicity extend in the sense that they also apply to this aspect of the choice of inquiry. This also means that, on our view, research projects can be deficient in terms of systematicity not only if they misconstrue the expected gain in understanding in a given field, but also if they misapprehend its significance in terms of what it would mean to society.

In such an admittedly restricted sense, both medical science and science in general are *moral enterprises*: their aim is to increase understanding that is significant in a rich sense, such that increasing understanding to deliberately decrease agency and autonomy does not count as true progress. Of course, as we have seen, there is an additional commitment in medical science to more or less directly contribute to the maintenance of health and the diagnosis, prevention, and treatment of disease. However, especially in light of the close connection between agency and autonomy, on the one hand, and health on the other, it is difficult to see why such an additional commitment would warrant an inference to a difference in kind. If this is correct, then (c) is false and the practical orientation of inquiry in medical science neither prevents it from being properly scientific nor renders inquiry in medical science different in kind.

4.7 Conclusion

Focusing on medical science, this chapter raised questions about the epistemic aim of inquiry. It put forward and defended the *Understanding Thesis*, according to which inquiry aims at *understanding*, while leaving the questions about what kinds of understanding characterize medicine for subsequent chapters. One major advantage is that comprehending progress in terms of increased understanding is shielded from some problems that accounts of progress measured in terms of increased knowledge face. The chapter acknowledged that while inquiry in medical science aims at understanding, it also displays a *practical orientation*, and provided arguments against the "purist" view defended by Pellegrino and others, which holds that due to such practical orientation, inquiry in medicine differs *in kind* from scientific inquiries. The chapter concluded that the differences in terms of aims, progress, and commitments are not decisive and definitively do not license inferring a difference in kind.

In offering support for these points, the chapter's findings make headway toward a more comprehensive account of the aim of medicine, to which Chapter 6 is dedicated. But they also feed into Chapter 5, which looks more closely at forms of understanding in medicine. The intimate link between understanding and increasing agential capacity that this chapter highlighted helps bring out that what constitutes adequate understanding is context sensitive, admits of degrees, and can take on different forms depending on the nature of the scientific field and the features of its subject matter. If so, then understanding in medicine might display unique characteristics that merit further investigation. Chapter 5 attempts to illuminate this issue.

CHAPTER 5

Understanding in Medicine

5.1 Introduction

Reflecting an interest in increasing the transparency of the world, we engage in epistemic endeavors that span from everyday, rudimentary inquiries to structured scientific inquiries. Inquiries in medical science, according to the *Understanding Thesis*, not only aim at furnishing truth or knowledge about some target question or phenomenon, but *understanding*, which can be seen as a distinct cognitive accomplishment. Due to the significant diversity that characterizes scientific endeavors, what qualifies as constituting proper understanding is to a certain degree context sensitive and can take on different forms depending on the nature of the scientific field and the features of its subject matter. If so, then we have at least some initial reasons for thinking that understanding within the context of medicine might differ in various ways from understanding in other areas such as physics or chemistry. A better comprehension of the nature of understanding in medicine merits sustained philosophical attention, and this chapter is dedicated to clarifying this matter.

Our starting point ensues from a number of points made in previous chapters. If inquiry in medical science has a scientific character (*Systematicity Thesis*) and aims at providing understanding (*Understanding Thesis*), then it seems safe to assume the working thesis that medicine aims to understand pathological conditions. In addition, because pathological conditions are in general detrimental to human agency, it seems plausible to think that the underlying motivation to understand them is to be able to intervene (i.e., cure, treat, prevent, predict, control) on them in a way that promotes our abilities as agents. But this initial picture is incomplete and leaves open major questions about (a) the particular nature of understanding in medicine and (b) how exactly medicine contributes to supporting human agency. While Chapter 6 will deal with (b), the chief task of this chapter is to shed light on (a) and hence to explicate understanding in medicine and some features that distinguish it from understanding in other scientific fields. Because the ambition of medical research is fused with medical practice, it is advantageous to approach the issue from the angle of clinical medicine.

The chapter falls into three parts. The first part describes in more detail what it means to understand something, distinguishes types of understanding, and links a type of understanding (i.e., objectual understanding) to explanations. The second part proceeds to investigate what understanding of a disease (i.e., biomedical understanding) requires by considering the case of scurvy from the history of medicine. The main hypothesis here is that grasping a correct mechanistic explanation of a condition is a necessary condition for biomedical understanding of that condition. The third part of the chapter argues that biomedical understanding is necessary, but not sufficient, for understanding in a clinical context (i.e., clinical understanding). The hypothesis is that clinical understanding combines biomedical understanding of a *disease* or pathological condition with a personal understanding of the patient with an *illness*. It will be shown that in many cases, clinical understanding necessitates adopting a particular second-personal stance and using cognitive resources *in addition* to those involved in biomedical understanding. The attempt to support this hypothesis will include revisiting the distinction between "understanding" and "explanation" familiar from debates concerning methodological principles in the humanities and social sciences.

Throughout the chapter, considering the various ways in which "knowledge" and "understanding" are used in everyday life will offer helpful guidance, but data about epistemic terminology from everyday parlance are unlikely to suffice for addressing substantial questions about the nature of understanding in the context of scientific inquiry. For example, everyday language not only often uses "understanding" and "knowledge" interchangeably, but it also permits utilizing "knowledge" in a manner that requires neither belief nor truth. For these reasons, consistent with the normative approach outlined in Chapter 1, the goal is not merely to report on findings of our analysis of ordinary language, but to "engineer" the relevant concepts in a manner that is suitable to the task and the subject matter at hand.

5.2 Knowledge and Types of Understanding

Epistemologists and philosophers of science have rediscovered understanding as a cognitive achievement that merits study on its own, and the rehabilitation of the notion of understanding is propelled by several factors

(Baumberger, Beisbart, and Brun 2017; Grimm 2021). First, some have argued that knowledge carries no distinct epistemic value above the sum of its proper parts (i.e., truth and justification), which makes it hard to maintain that knowledge merits the attention that it has received in epistemology (Kvanvig 2003; Pritchard 2010). Second, the somewhat myopic focus on knowledge tended to ignore what motivates and bestows value on our inquiries from an epistemic perspective, which is, in general terms, to understand the world we inhabit and to render it more transparent to us.

"Knowing" and "understanding" are closely related cognitive achievements, occur in similar linguistic forms (one can know-how, know-that, and know-who, just as one can understand-how, understand-what, and understand-who), and are often used interchangeably (Kvanvig 2009; Hannon 2019, ch. 9). We operate with different uses of understanding, and we regularly claim to understand computers, languages, other human beings, symbols, why and how certain events occurred, and so on. Disregarding, for instance, linguistic understanding (e.g., "I understand the meaning of 'tool'"), propositional understanding ("S understands that he needs to pass the exam"), and nonexplanatory understanding (e.g., "I understand who my friends are"), the most relevant types of understanding for our context are:

1. *Explanatory understanding*: "S understands why X is the case"
2. *Objectual understanding*: "S understands X" (e.g., object, subject matter)[1]
3. *Practical understanding:* "S understands how to X"

While the first two types of understanding each mirror a type of knowledge, in each case knowledge is not sufficient for understanding. In the case of explanatory understanding, S might know the cause of X, but S can only be said to exhibit understanding if S grasps how the cause brings about X. Recall Pritchard's (2010) example in which a young boy comes to know by testimony from a reliable source that the house burned down due to faulty electrical wiring. The boy attains (causal) knowledge, but in lacking some idea of *how* faulty wiring might bring about a fire, he does not attain the relevant piece of explanatory understanding.

In the case of objectual understanding, the situation is similar. This is the sort of understanding that one can acquire of a domain or subject matter (Kvanvig 2003, 191; 2009) and it is usually ascribed by means of the verb "understands" followed by a noun ("S understands scurvy").

[1] Objectual understanding can also have as its object a theory, but this might be a special case, as it is often a means to achieve objectual understanding of subject matters, processes, etc.

In such cases, saying that S understands X is attributing to S a more profound penetration of the target, an intimate epistemic acquaintance that outstrips knowledge of individual propositions (Riggs 2003; Strevens 2017). S can have acquired knowledge of countless isolated bits of information about X by testimony, but this would not be sufficient to rise to the level of understanding. Objectual understanding displays "multiple gradability" (Bengson 2017), such that it can always be deeper or richer along various dimensions.

Practical understanding ("understanding-how") has been less prominent in the debates in epistemology. It is sometimes contrasted with "theoretical understanding" (Lipton 2009), although some argue that they possess a common underlying nature (Bengson 2017). The paradigm case of practical understanding is a skillful activity that differs from reflexive behaviors and underscores that practical understanding is not about explanations but about certain bodily or mental abilities. Practical understanding is in this sense not explanatory (Khalifa 2013) and builds on nonpropositional knowledge that is not vulnerable to Gettier-style defeaters.

5.3 Grasping Explanations

According to a widely accepted view, explanatory understanding and objectual understanding involve an additional cognitive achievement that distinguishes understanding from knowledge, and many conceive of this as a kind of "grasping" (see, e.g., de Regt and Dieks 2005; Grimm 2006; 2014; 2016; Elgin 2007; 2017; de Regt 2009; Newman 2012; Khalifa 2013; Hills 2016; Strevens 2017). As Jonathan Kvanvig (2003, 192) puts it, "understanding requires the grasping of explanatory and other coherence-making relationships in a large and comprehensive body of information." There is no consensus on what "grasping" precisely consists of, and there is an unfortunate tendency to use the term in a largely metaphorical way (Hannon 2019), even though many agree with Michael Strevens (2017, 41) that "to grasp a fact is like knowing a fact, but it involves a more intimate epistemic acquaintance with the state of affairs in question." To clarify these matters, we may make two points about (a) what grasping is and (b) what is grasped in understanding.

As to (a), we may comprehend grasping as some kind of *cognitive command* or cognitive control (Hills 2016) that we acquire upon the exercise of our epistemic agency in assembling information. This in turn allows us to make transparent and represent conceptual and explanatory connections between parts and processes. When one possesses cognitive command, one is able to mentally map a relational assembly that allows one to exploit the information in some way. Consider again the boy who has attained causal knowledge. When we say that he lacks understanding, what best describes the most relevant difference is that he lacks cognitive command of explanatory dependency relations with respect to the event. This renders him unable to reach correct conclusions about similar cases in slightly different circumstances by engaging in counterfactual reasoning.

When speaking of grasping as some kind of ability of cognitive command, it is important to avoid reducing the *subjective aspect* of understanding to a *phenomenological aspect* (i.e., a so-called aha feeling, some sort of mental click). Good explanations may fail to yield a sense of understanding, while poor explanations might produce a bias in scientists for thinking that the correct explanation has been found (Trout 2002). While there are obvious problems with assessing the quality of explanations with recourse to the sense of understanding, it is not clear whether this can be avoided entirely (Wilkenfeld 2013), and the sense of understanding may be mediated simultaneously by subjective experience and still function as a reliable source of knowledge (Lipton 2009).

As to (b), at least in most cases in the context of the sciences, what are grasped are *explanations*. While it is intuitive to think that anything deserving the label "explanation" ought to be capable of clarifying something previously unclear, some of the debates have comprehended understanding as something that is produced by having an explanation. Carl G. Hempel and Paul Oppenheim (1948, 145) argued that understanding is produced by deductive-nomological explanations while causal-mechanistic accounts connect explanation to identifying causal mechanisms responsible for the target phenomenon (Salmon 1984, 132; Machamer, Darden, and Craver 2000). Taking this further, others have supposed that understanding is *the point* of explanation: science wants to explain some target phenomenon X *because* we want to understand X (Lipton 2001). Whether or not this stronger point is correct, it is worth noting that the discussions sometimes suggest that understanding lies in the objective relation of subsumption under laws or the identification of causal mechanisms. For our purposes, it is important to bear in mind that understanding requires some additional subjective aspect linked to representing those relations.

5.4 Objectual Understanding

Much of the literature distinguishes between objectual and explanatory understanding and highlights differences between their objects, the cognitive efforts they involve, and what is distinctively valuable about them (Kvanvig 2003; 2009; Carter and Gordon 2014; Baumberger 2019; Hannon 2019). However, some think that, ultimately, objectual understanding is reducible to having some sufficiently large amount of relevant explanatory understanding (see, e.g., Khalifa 2017, ch. 4). The idea is, roughly, that explanatory understanding takes as its object a state of affairs (e.g., that the occurrence of scurvy has rapidly increased), while objectual understanding takes as its object a subject matter (e.g., scurvy), which itself is nothing more than a composite of states of affairs (Grimm 2011). This is a complex issue, and it is not possible to adequately cover the details of this debate within the confines of this chapter. However, there are at least four reasons for maintaining a distinction between objectual and explanatory understanding.

First, some have argued for the general view that attaining objectual understanding is the goal of inquiry (Kvanvig 2013; Carter and Gordon 2014). Objectual understanding is what satisfies the desire to comprehend a subject matter, and attaining it legitimately closes the investigation into the subject. At least in the context of medicine, objectual understanding appears to more adequately describe the ultimate goal of inquiry. Of course, we want to understand why a condition like scurvy arises, but we also want to understand why scurvy takes on the form it does, how it is correlated with other conditions, what effects it has on the mind, how its impact varies across individuals, how we can systematically describe and classify its signs and symptoms, etc. In short, medicine does not merely aim to obtain explanatory understanding of the features of scurvy (e.g., why various biochemical reactions occur, why the ingestion of citrus fruits mitigates the symptoms). Instead, the goal is to systematically *understand scurvy*, to attain some coherent completeness with respect to knowledge, classifications, and taxonomies, even if single inquiries cannot take on such a large task. Such systematic understanding of a subject matter also indicates that the focus of objectual understanding is broader than that of explanatory understanding (Hannon 2021).

Of course, one might agree that objectual understanding better captures the aims of inquiry, and that attaining it requires having a sufficient degree of explanatory understanding, but still insist that objectual understanding is reducible to explanatory understanding. Against such a view, Kvanvig (2009) argues that objectual understanding cannot be reduced to the latter because it is possible (e.g., in indeterministic systems) to attain some degree of objectual understanding where explanatory relations do not exist.

In such cases, while explanatory understanding is lacking, one can still attain objectual understanding by grasping other structural relationships such as probabilistic relationships. While some of the relevant examples and the claim of irreducibility might be disputed (see Khalifa 2013), for the purposes of this chapter we may note that the type of inquiry associated with medicine described earlier involves aspects that are not straightforwardly explanatory. For example, the classificatory efforts with respect to various diseases like scurvy and the taxonomy of different forms of scurvy *enhance* our objectual understanding of scurvy even if classifications in themselves do not enable us to explain facts about scurvy.[2]

There is perhaps also a relevant difference in the state of understanding. When we say that S understands scurvy, we attribute to S some significant level of cognitive command of scurvy, reflecting that she understands *how* and *why* various elements and aspects of what is explanatory with respect to scurvy hang together. S's epistemic commitments relevant to scurvy are interrelated in a coherent network, and she has a firm grasp of dependency relationships between a large number of items of information. Moreover, objectual understanding says something about *how* S holds an understanding of the relevant explanatory and other structural relationships, namely in a *systematic* fashion, offering her cognitive command over a body of interlinked information that enables further cognition and action with respect to the phenomenon. For example, given her objectual understanding, we might expect S to have a good idea of what explanatory understanding would be worth seeking in order to improve her understanding of scurvy. Such expectations would be unjustified if S's objectual understanding of scurvy was nothing more than a fragmented collection of explanatory understandings.

Another reason to maintain this distinction is that it conserves the intuition that the factivity condition of objectual understanding is less demanding than that of explanatory understanding, such that the former is less vulnerable to peripheral falsehoods (see, e.g., Kvanvig 2009; Baumberger et al. 2017; Elgin 2017). Moreover, as Baumberger et al. (2017) point out, the two forms of understanding can also be distinguished in terms of the means by which they are achieved. That said, perhaps we may remain agnostic about the question whether objectual understanding is reducible to explanatory understanding, but preserve the difference due to pragmatic reasons. As Michael Stuart (2017, 529) points

[2] For a general discussion of taxonomies and objectual understanding that makes a similar point, see Gijsbers (2013).

out, even if different forms of understanding were mutually reducible, "we would still want to keep them apart since there are different ways of obtaining each type of understanding and different ways of determining when each has been achieved. This is particularly clear in the scientific context."

5.5 Understanding Disease

The account of understanding offered here allows us to acknowledge that what constitutes suitable understanding in the different scientific fields might be different, in part, because what qualifies as a suitable explanation depends on the disciplinary context and the relevant research questions (for a discussion, see Strevens 2010). For example, understanding phenomena can involve using simulations based on mathematical models, finding robust patterns in large databases, direct intervention (e.g., manipulation of components in the case of biological entities), and so on. But even within the same branch of science, what constitutes understanding can be subject to change. For example, while Lord Kelvin maintained that scientific understanding in physics is achieved by developing mechanical models of physical phenomena, developments in physics such as the emergence of quantum theory undermined this idea (De Regt, Leonelli, and Eigner 2009).

Nonetheless, the account offered here provides the basis for exploring what objectual understanding of a disease (i.e., biomedical understanding) requires in the context of medicine. For this, we will now consider an example from the history of medicine.

5.5.1 Understanding Scurvy

Scurvy, a disease of malnutrition that we now know is caused by vitamin C deficiency, killed large numbers of sailors between the sixteenth and nineteenth centuries, and the default assumption of shipowners was that half the sailors on any major voyage would die from it. Although the effectiveness of a diet involving fresh fruits, especially lemons, as a prophylactic was already known to sailors in the early seventeenth century, physicians stuck to the diagnosis of "humoral imbalance" and recommended bloodletting, salt-water-induced vomiting, and the consumption of "fizzy drinks" to stimulate digestion.

In 1747, the naval surgeon James Lind conducted a landmark clinical trial involving twelve participants suffering from advanced scurvy.

Persuaded that in order to understand disease and develop a cure treatments had to be assessed in similar patients and simultaneously (Lind 1772, 149–52), Lind divided the patients into six groups. One group was given cider, one received elixir of vitriol, one vinegar, one salt water, one a laxative, and one oranges and lemons. In current terms, Lind's undertaking was a six-armed, non-controlled comparative trial that concurrently studied commonly used scurvy treatments. Holding all other circumstances constant, Lind supplied groups of randomly chosen individuals with different alleged remedies and observed whether or not a significant difference was produced in the course of the scurvy. The result of Lind's study was relatively unambiguous: the sailors receiving oranges and lemons were cured, but not the others.

Lind's experimental intervention advanced the explanatory understanding of scurvy by distinguishing variables that make a difference for the occurrence of scurvy from those that are simply correlated with it (e.g., sea travel, crowded living spaces, hard work). To shed light on how Lind proceeded and how causal relationships are comprehended in the sciences, we may return to accounts of causation that subscribe to the view that a cause must make a difference to its effects (see Woodward 2003; 2010; 2015). A causal relationship holding between variables X and Y is a matter of counterfactual dependence between X and Y. Claiming that X causes Y (represented as X → Y) means that there is a possible manipulation of some value of X, which, under the appropriate conditions, will change the value of Y or its probability distribution (Woodward 2003, 40; 2010, 290). A causal connection holding between X and Y can be explicated in terms of the results of an ideal experimental intervention I on X with respect to Y. I is an intervention variable for X with respect to Y if and only if I meets four conditions (Woodward 2003, 98):

(1) I causes X.

(2) I acts as a switch for all the other variables that cause X. That is, certain values of I are such that when I attains those values, X ceases to depend on the values of other variables that cause X and instead depends only on the value taken by I.

(3) Any directed path from I to Y goes through X. . . .

(4) I is (statistically) independent of any variable Z that causes Y and that is on a directed path that does not go through X.

This account offers a way to probe an alleged causal relation by experimental manipulation, consisting of an intervention that manipulates the putative cause and observes whether an effect results. It does not merely offer a tool to distinguish correlation and causation: it also fits a fundamental pragmatic objective, which is to causally intervene to treat or prevent diseases (Kendler and Campbell 2009).

Returning to Lind's experimental intervention, we see that it meets condition (1), because the intervention determined the level of lemon and orange consumption. Condition (2) is met because only Lind's choice to assign a particular diet to a group determined whether the group had a high consumption of lemon and orange. Condition (3) is met because if assigning people in various groups influenced their scurvy, then it did so only by way of the elevated consumption of the relevant substance. Finally, condition (4) is met so long as the intervention was a randomized experiment in which fixing the value of the consumption of lemons and oranges was independent of other variables that might have influenced the course of scurvy.

5.5.2 Some Limitations: The Lack of a (Correct) Mechanism

Lind's contribution to understanding scurvy was of course enormously important: unlike the causal relationship between the ingestion of oranges and lemons and scurvy, the correlational relationship between the ingestion of nutmeg, vinegar, or salt water and scurvy is not exploitable for manipulation and control. Even so, his comprehension of the causal relation of oranges and lemons on scurvy was relatively rudimentary, as he had no knowledge of two important dimensions of the causal relationship, namely *stability* and *specificity* (for a general discussion, see Woodward 2010).

The greater or lesser stability of a causal relationship depends on the number of background circumstances in which it occurs. The causal relationship between oranges and lemons (OL) and scurvy (S) is relatively stable if the counterfactual dependence holds under a wide variety of background circumstances. Moreover, OL → S can be more or less specific, referring to the grain level of counterfactual dependencies between OL and S. OL → S is specific if the counterfactual dependencies holding between OL and S are fine-grained, in which case the manipulation of OL enables more precise control over the value of S. X → Y has a high specificity when intervening on X enables modulating the state of Y in a fine-grained manner. Conversely, switch-like causation has a low grade of causal specificity. Clearly, Lind's understanding of scurvy would have been more profound had he attained some grasp of the stability and causal specificity of the relationship between OL and S.

While Lind's rudimentary comprehension of some causal dependencies allowed him to manipulate the condition, it would be odd to claim that he understood scurvy in the sense of objectual understanding. One could say that he made important steps toward achieving objectual understanding, or perhaps even that he attained some degree of objectual understanding, but it would be excessive to claim that he had obtained objectual understanding of scurvy in any significant sense. Importantly, Lind did not know whether scurvy was somehow caused by a diet lacking lemons and oranges, which also meant that he could not refute other explanations of why scurvy occurs (e.g., bad air or crowding that somehow disturbs humoral balance, which can be relieved by ingesting lemons and oranges). Of course, he could have used the same procedure to determine whether the lack of lemons and oranges causes or increases the risk for scurvy. He could have designed an experiment in which he randomly intervened on individuals in a given population exposing them to the lack of citrus fruits and observed if they subsequently suffered from an increased incidence of scurvy. Still, objectual understanding of scurvy requires something else that allows one to "trace" the causal process (Steel 2008) and that helps to piece causal information together and grasp coherence-making relationships. It requires some degree of explanatory understanding, which could be attained by *identifying the mechanism* responsible for the causal connection between the two variables.

Roughly, a mechanism M for a phenomenon P consists of parts, the activities of which are organized in a way that they are responsible for P (Glennan, Illari, and Weber 2022). Outlining the spatiotemporal and hierarchical organization of mechanisms (e.g., biochemical pathways) plays a key explanatory role in the biomedical sciences by shedding light on the proper function of features of the body and the emergence and progress of diseases (Williamson 2019). Placing additional emphasis on mechanisms, some hold that establishing the claim that $X \rightarrow Y$ not only requires difference-making evidence (e.g., the kind of evidence that Lind had gained), but also evidence of a satisfactorily delineated mechanism constituted by entities (e.g., proteins) and activities (e.g., protein expression) linking X and Y (Russo and Williamson 2007). While many have argued that this thesis is too strong and that causal claims can be accepted on the basis of clinical studies alone (e.g., Broadbent 2011; Howick 2011), a weaker thesis is acceptable. According to the weaker thesis, evidence of a mechanism in conjunction with evidence of difference-making helps increase confidence that the observed correlation between X and Y is not spurious and that changes in Y can be attributed to the experimental intervention on X and not to confounding (Illari 2011; Williamson 2019). Similarly, evidence of the absence of a possible mechanism linking X and Y decreases confidence that there is a causal relationship. For example, in an evaluation of whether mobile phone usage can cause cancer, a significant correlation was found between certain forms of cancers and high levels of call time. However, because it was found unlikely that there is a mechanism connecting the purported cause and effect, the conclusion was that the result is best explained as due to error or chance (Williamson 2019).

While evidence of mechanisms may not be necessary for establishing causal claims, attaining understanding seems to require both difference-making evidence and some evidence of mechanism. Lind's study helps establish a difference-making relationship, but because it does not identify a correct mechanism linking cause and effect such that citrus fruits act to prevent scurvy, it fails to offer the kind of explanatory understanding that – together with bits of knowledge – could amount to biomedical understanding.

Lind made attempts to reach this stage of understanding. He tried to offer an account of the relevant mechanism by which scurvy is produced and why lemons and oranges had a positive effect on it. However, his account relied on the humoral theory of disease. Roughly, he claimed that (a) perspiration through the skin is vital for the balance of the humors, (b) scurvy involved a blockage of the pores caused by damp air, and (c) lemons and oranges had the capacity to dissolve it. As Leen De Vreese (2008, 22) puts it, "the conceptual framework which could have provided him understanding of the real mechanisms leading from such a nutritional deficiency to the development of the disease was entirely lacking." In this manner, De Vreese (2008, 15) notes that the wrong account of the mechanisms has led to *mis*understanding, only seemingly enhancing explanatory coherence. This flawed account of the mechanisms prevented answering a variety of what-if-things-had-been-different questions, and anticipation of the effects of certain conceivable interventions. As understanding is a success term that involves some degree of factivity, one might claim that Lind's inquiry counts as increasing or making steps toward understanding by way of uncovering a causal relationship, but not that he attained explanatory or objectual understanding.

The failure to identify the mechanism responsible for the causal connection is also the reason why Lind encountered a major setback later in his career. He began treating patients with concentrated lemon juice that had been heated, which destroyed much of the vitamin C. Unwavering in his commitment to humoral theory, Lind conducted no tests to compare his boiled concentrates with fresh fruits and ultimately returned to bloodletting (Wootton 2006).

5.5.3 *Biomedical Understanding and Mechanistic Explanation*

Before exploring additional gains in understanding scurvy, it is helpful to add some clarifications about mechanistic explanations that are prevalent in the biological and behavioral sciences. An etiological mechanistic explanation typically comprehends phenomena in terms of their being caused by a mechanism, defined as "a structure performing a function in virtue of its component parts, component operations, and their organization" (Bechtel and Abrahamsen 2005, 423).[3] In contrast, a constitutive mechanistic explanation advances understanding by recourse to the behaviors and organization of component entities of the underlying mechanism, which stand in a constitutive relationship to the phenomenon.

Explanations in medicine are often modelled on biological explanations, which are most frequently defined as mechanistic explanations involving biochemical mechanisms (Thagard 2003; 2005; Kaplan and Craver 2011; Darrason 2018). For example, Paul Thagard defines medical explanations as identifying the "mechanisms whose proper and improper functioning generate the states and symptoms of a disease" (Thagard 2005, 59). Disease is thus the product of altered biological mechanisms, or in some cases perhaps the product of autonomous pathological mechanisms (Nervi 2010), but explanations in medicine are typically mechanistic such that they explain a disease by localizing and disclosing the spatiotemporal organization of a mechanism that produces its symptoms. As Marie Darrason (2018, 149) puts it, "most medical explanations are considered mechanistic explanations: in order to explain a disease, you need to localize and decompose the mechanism that produces the disease symptoms." Mechanistic explanations of diseases have certain advantages: identifying the mechanisms responsible for the disease permits going beyond pure phenotypic characterization of disease and helps illuminate what restoring the dysfunctional mechanism would require.

To further support the idea that biomedical understanding requires grasping mechanistic explanation, we may now continue our exploration of the history of scurvy. A decisive step toward understanding occurred at the beginning of the twentieth century, when Norwegian researchers Axel

[3] See Illari and Williamson (2012) for different characterizations.

Holst and Theodor Frølich were developing an animal model for "ship beriberi," which resembled scurvy in a number of ways. They suspected a nutritional deficiency and tested the idea on guinea pigs, which, incidentally, are among the few mammals unable to endogenously synthesize ascorbic acid. Holst and Frølich found that guinea pigs on a diet of grains developed scurvy-like symptoms, and their autopsies showed signs of scurvy but not beriberi. Subsequently, their studies indicated that symptoms could be neutralized by putting the animals on a diet of fresh foods (apples, cabbage, potatoes, and lemon juice), and they proposed that these contained a special substance that mediates the causal relationship between nutrition and scurvy (Combs and McClung 2016, 18).

Holst and Frølich's findings supported the idea of a nutritional deficiency causing scurvy, but the crucial factor was only discovered two decades later when Albert Szent-Györgyi isolated a molecule that he termed "hexuronic acid." Together with Joseph L. Svirbely, Szent-Györgyi conducted further experiments using guinea pigs. One group received boiled food, which destroyed the vitamin C, while the other received food supplemented with hexuronic acid. The animals in the first group developed scurvy while those in the second group did not. Svirbely and Szent-Györgyi argued that hexuronic acid was responsible for protection from scurvy in the second group and renamed it "ascorbic acid" to highlight its anti-scurvy effects. Ascorbic acid eventually became known as vitamin C (Carpenter 2012).

Although the causal agent was found, there was still no comprehension of how the elements of scurvy are configured. What is it that binds together putrid gums, spots, fatigue, and joint pain? Are these symptoms parts of scurvy or are they caused by scurvy? A final breakthrough in this regard was the discovery of the *metabolic mechanism* that is responsible for the synthesis of collagen, which requires vitamin C for its functioning. More precisely, vitamin C is a cofactor for two enzymes (prolylhydroxylase and lysylhydroxylase) that are responsible for the hydroxylation of collagen. These enzymes require vitamin C to be present as a cofactor, and deficiencies of vitamin C, such as in scurvy, can cause defects in collagen. This explanation identifies the mechanisms responsible for the normal functioning of collagen hydroxylation *and* a way in which vitamin C deficiency is a factor that can interfere with it.

With a mechanistic explanation of normal and altered collagen synthesis as constituted by the configuration and activities of component entities – entities in the mechanism relevant to its operation – researchers took a leap toward objectual understanding, that is, toward being able to construct a web of relational networks that include correlations, causes, and mechanisms. The identification of a fine-grained mechanism increased explanatory power, allowing for more what-if-things-were-different questions to be answered and doing so in a more precise manner. The mechanism explains not just what causes scurvy, but also why some tissues such as skin, gums, and bones with a higher concentration of collagen are more disposed to be affected. The seemingly disparate symptoms now stand out as a coherent whole: they do not stand in a causal relationship to each other, but are connected by a common cause.

5.5.4 *Mechanisms and Two Types of Dependence*

One might agree that grasping a mechanistic explanation is required for objectual understanding in biomedicine, but insist that the information about mechanisms can be reduced to information about fine-grained causal relations.[4] While some stress that information about mechanisms involves more than that, as mechanisms are truth-makers for causal claims (Waskan 2011), for our purposes we need not take sides on the details (for a recent discussion, see Craver, Glennan, and Povich 2021). Instead, we merely point out that while both causal and mechanistic explanations map networks of counterfactual dependence, it is often helpful to keep separate two difference-making relationships. In a mechanism, there is a horizontal (causal) dimension and a vertical (part-whole) dimension, such that the relations among the components are *causal*, while the relationship between individual components and the phenomenon is *constitutive* (Craver 2007; Craver and Bechtel 2007; Glennan 2017; Craver and Tabery 2019).

The difference is that in a causal relationship between X and Y, dependence is *asymmetrical* (If X causes Y, then Y depends on X but not vice versa), and X and Y are mereologically independent entities such that X temporally precedes Y. However, if X constitutes Y, then their relationship of dependence is *symmetrical* (Y depends on X and vice versa), and X and Y are spatiotemporally overlapping entities. Due to such differences, many maintain that the relations of dependence supporting constitutive mechanistic explanations of activities of wholes using activities of components are not causal (Gillett 2020), but they can make sense of how wholes

[4] As Woodward (2004, 60) puts it, "I certainly don't dispute the importance of information about intervening mechanisms, but see this as more information about additional, fine-grained patterns of counterfactual dependence, rather than as information that dispenses with counterfactuals in favour of something else."

have their causal capacities by appealing to their components and their organization (Ylikoski 2013).

These distinctions in terms of difference-making are important for biomedical understanding, and scientists have established experimental methods to distinguish genuine components from causal factors. To discover causal relationships, scientists unidirectionally intervene to manipulate X and thereby change Y, but given the symmetrical nature of constitutive relations, this is clearly not sufficient. On the basis of an examination of explanations in the biological sciences, Carl Craver (2007) has put forward what he refers to as the *mutual manipulability* approach to constitutive relevance. The main idea is that the parts and their activities are constituents of a phenomenon if the relevant interventions uncover mutual difference-making. So X is constitutively relevant to Y if the relata stand in a part-whole relationship and if they are mutually manipulable, such that there is a possible ideal intervention on Y under which X is altered and vice versa (Craver 2007, 152–3; for recent clarifications, see Craver et al. 2021).

Constitutive relations are established by executing interventions on a phenomenon in a top-down manner and on the parts in a bottom-up manner. Bottom-up experiments intervene on putative lower-level components (often by boosting or reducing their activity) while tracking changes in the phenomenon at a higher level. In contrast, top-down experiments intervene on the level of the overall phenomenon while tracking changes on the lower level of putative components. Although determining the adequate combination of bottom-up and top-down interventions will vary from case to case, demonstrating constitutive relevance necessitates performing both top-down and bottom-up experiments (Craver and Darden 2013).

In all, mechanisms matter for achieving some acceptable level of systematic, biomedical understanding of a condition. Mechanistic explanation helps map a rich network of counterfactual dependence that enables increasing opportunities for intervention.

5.5.5 *Summing Up*

Thus far, the chapter has offered support for the view that in the medical context, grasping a mechanistic explanation of a disease is a necessary condition for attaining an objectual understanding of it. In Section 5.6, we turn our attention to the question of what understanding is in a clinical context and we shall see that biomedical understanding as developed here is required for clinical understanding. But before proceeding, we should note that the approach chosen here has some rather serious limitations. The chapter has only considered a single case, so, without further arguments, the conclusion is not very well supported. However, scurvy is commonly discussed as a prototypical disease, so there are at least some reasons for thinking that our findings will also apply to a wide range of other diseases. Moreover, it appears that the idea that objectual understanding in the context of medicine requires grasping a mechanistic explanation also applies in standard clinical settings. In regular clinical medicine it seems difficult to think of cases in which physicians display objectual understanding of a condition while having no grasp of a mechanistic explanation with some reasonable level of detail. Being aware of a causal connection between headaches and paracetamol enables one to intervene on headaches, but understanding the difference-making relationship does not suffice for objectual understanding of headaches unless one has some grasp of the specific mechanism connecting them, which also explains why intervening on one variable makes a difference to the value of the other. Of course, our understanding of a number of conditions is still very fragmentary, and in those cases it might not be unwarranted to attribute objectual understanding. Moreover, what constitutes a reasonable level of detail will depend on its relevance for diagnosis, treatment, and prognosis.

5.6 Clinical Understanding

While grasping a mechanistic explanation is a necessary condition for biomedical understanding, in standard clinical situations biomedical understanding has to be adequately contextualized and supplemented. In order to make an accurate diagnosis, the medical professional initiates a systematic inquiry to gather an interconnected body of information. To achieve this in an optimal fashion, biomedical understanding will, in many cases, need to be complemented by a participatory, subjectively involved form of *personal understanding,* which necessitates adopting a particular second-personal stance and using cognitive resources *in addition* to those involved in biomedical understanding. This requires going beyond understanding *disease* (i.e., how prototypes of diseases manifest themselves in unique individuals) to understanding *illness* in its specificity that reflects the individual's distinctive predicament. In turn, at least in some cases, biomedical understanding will be required to assist personal understanding, indicating that biomedical and personal understanding can be entangled.

The attempt to clarify these aspects and identify unique features of clinical understanding takes us back to the juxtaposition of explanation and understanding, which has played a key role in attempts to locate medicine in relation to the biological sciences, on the one hand, and the distinctively human sciences on the other (see, e.g., Wartofsky and Zaner 1980). Contrasting explanation (*erklären*) with understanding (*verstehen*), influential figures in the nineteenth century, like Johann G. B. Droysen and Wilhelm Dilthey, argued that this distinction marks a methodological division between the humanities and the social sciences, which are essentially aimed at understanding, and the natural sciences that are essentially aimed at explanation. On this view, explaining (*erklären*) designates primarily causal explanations in the natural sciences (primarily physics) dealing with phenomena that are amenable to explanation in terms of laws (or law-like regularities).

In contrast, inquiries in the social sciences and humanities aim at understanding (*verstehen*) and follow a different path, because comprehending human behavior requires capturing the *meaning* the events have for the subjects (Taylor 1971). Thus, understanding often involves making sense of other people's mental processes, and it refers to a form of comprehension that we can acquire of human cognition, psychological states, action, artifacts, and institutions, but not of the kinds of entities and processes that the natural sciences typically deal with.

This brief sketch cannot do justice to the depth of the different positions, but it is sufficient to comprehend why many have thought that medicine stands at the center of this methodological distinction, integrating both *erklären* and *verstehen*. While there are reasons to attempt to identify the specific scientific character of medicine by recourse to this juxtaposition, there are also several reasons why approaches choosing this path have not been very productive.

First, the juxtaposition builds on faulty assumptions about explanation in the sciences. For example, it is often based on the assumption that explanation in the natural sciences is aptly characterized by the D-N model, on which explanations involve at least one law plus the initial conditions, from which explanations emerge much like logical proofs. This, however, does not sit well with the fact that explanations in a range of scientific domains like biology do not necessarily invoke laws (Bechtel and Abrahamsen 2005). If one abandons the idea that causation requires laws and accepts that it only requires counterfactual dependencies, then the strict juxtaposition becomes untenable.

Second, any account that imposes such a strict juxtaposition between explanation and understanding is forced to accept the view that inquiries in the natural sciences never achieve understanding of the subject they study, while the humanities and social sciences never explain anything (Stueber 2012). This not only sounds intuitively implausible to contemporary ears, it is also inconsistent with the account of science as systematic inquiry.

Consequently, present work in epistemology and the philosophy of science challenges this juxtaposition (see, e.g., Khalifa 2019), and recent accounts of understanding encompass the human and natural sciences. However, with respect to medicine specifically, rejecting such a juxtaposition should not lead us to lose sight of important aspects that can be illuminated along the understanding vs. explanation distinction. Even though the juxtaposition of explanation and understanding is misguided, keeping it in mind might help us remain sensitive to the intuitively plausible idea that when the subjects of inquiry are human beings instead of organs, tissues, or proteins, understanding and explanation may take on different forms and require different methods and cognitive efforts. Operating with a context-sensitive notion of understanding, we are able to acknowledge that there is something distinct about understanding human beings, without having to impose a stark methodological division between the human/social sciences and the natural sciences.

Applying these reflections to medicine, Sections 5.6.1 and 5.6.2 will seek support for the thesis that clinical medicine involves understanding in the sense of *verstehen*, because comprehending the condition that a patient seeks help with often requires capturing the *meaning* that health-related events have for them. Put differently, when seeking to make an accurate diagnosis and devise a treatment plan, successful application of the biomedical understanding of the disease often requires understanding the *illness*. Roughly speaking, illness includes the patient's perspective on their ill health, its perceived origin and significance, and the meaning the patient gives to that experience, all of which are profoundly influenced by sociocultural background and personality traits. Understanding illness is comprehending subjective aspects of how the disease is experienced, including the patterns of emotions, reasoning, and actions that it is associated with (for a discussion, see Hofmann 2016). Such understanding is essential, especially in cases in which the therapeutic encounter not only aims to classify and treat disease, but also to offer comfort and care. However, understanding illness necessitates some form of *personal understanding* that deploys cognitive resources in addition to those involved in understanding required for explaining and predicting the behavior of purely biological processes. Such personal understanding can be vital to the development of a therapeutic relationship and to practicing medicine effectively. To see how, we start by considering some characteristics of the medical interview.

5.6.1 *Personal Understanding: Clinical Empathy*

Usually prompted by patients requesting help with specific health problems, medical interviews assess current risk factors in part by collecting relevant information about the patient's family history, past medical history, and social history (e.g., occupation, marital status). When gathering evidence, the interviewer solicits information that permits more fitting hypotheses and often prompts further questions. The information received is evaluated for reliability, comprehensiveness, and significance to the patient's problem and it is examined for symptom complexes and clues about possible underlying conditions that might explain the patient's complaints. In spite of the advances in laboratory testing, the medical interview continues to retain an important role: an accurate history alone is often sufficient for a diagnosis, and it is essential for focusing the scope of further diagnostic examination.

What needs to be understood in the clinical interview is not merely how prototypes of diseases described in medical textbooks manifest themselves in particular individuals. In other words, what needs to be understood is *illness*, not disease. As Peter Lichstein (1990) puts it, "patients rarely report their symptoms in an organized and logical fashion comparable to the descriptions of disease in medical texts. In fact, patients complain of illness or sickness rather than stating their problems in terms of the pathophysiologic categories of disease." Accordingly, one task in the medical interview is to complement biomedical understanding of disease with understanding the illness in its specificity, which, in turn, requires minimal personal understanding. To illustrate how this transpires, it is helpful to turn our attention to how interviewing strikes a balance between leading the interaction and assisting the patient's spontaneous report.

Patients do not present symptom complexes in an organized fashion, and no matter how experienced and skilled clinicians are, they cannot simply "extract" a history from a patient (Reiser and Schroder 1980). To acquire comprehensive information that cannot be obtained from other sources (i.e., *what* the patient says and *how* the patient says it), attention has to be paid to verbal and nonverbal aspects of the patient's behavior during the interview, especially given that patients typically meet the clinician in a situation of heightened anxiety and vulnerability. Facial expressions, posture, gestures, along with abrupt changes in topic and evasion of certain issues, may constitute reactions to illness or indicate concerns that are not directly expressed. For example, a patient might state that she is feeling excellent, but the physician might detect distress prompting further questioning or examination. In such cases, information that physicians need to register and pursue is often only nonverbally hinted at (Suchman et al. 1997; Halpern 2014).

What emerges is that in the context of clinical medicine, efficiently deploying biomedical understanding of pathological conditions often requires some degree of personal understanding. To see what this amounts to, we may start by drawing attention to certain aspects of clinical communication that are often described under the heading "clinical empathy." According to a common definition, clinical empathy is "the ability to understand the patient's situation, perspective and feelings, and to communicate that understanding to the patient" (Coulehan et al. 2001). Such ability to understand, which we describe here as a minimal personal understanding, is expressed in "active listening" techniques that show involvement, reflect the patient's message (e.g., by using verbal paraphrasing), and encourage patients to elaborate on key symptoms or experiences.

This is obviously a complex topic, and definitions of what constitutes clinical empathy have definitive weaknesses, but in this context we may suffice with highlighting that personal understanding can be important for obtaining a comprehensive history and for making the correct assessment of the patient's condition. One reason is that the quality of the information collected during the clinical interview depends on the quality of the connection that develops between physician and patient. A good rapport is best achieved by interacting with the patient in an attuned manner, which requires the kind of ability to understand characterized by "clinical empathy." For further support, it is helpful to consider studies reporting that patient-perceived empathy is associated with reduction in severity and duration of symptoms and with positive clinical outcomes in the common cold and diabetes (Hojat et al. 2011; Rakel et al. 2011). This effect can in part be explained by patients being more open about their intimidating symptoms and psychosocial concerns to a physician displaying clinical empathy, which leads to a more accurate diagnosis and better compliance with proposed therapies (Neumann et al. 2009). Observational studies of patient-physician interactions show that patients' decisions to either omit or disclose fuller histories and anxiety-provoking symptoms depends on whether they sense that the physicians are emotionally attuned to them (Halpern 2014). Based on patient interviews with primary care physicians, other studies report that physicians who acknowledge emotive cues and probe for further information based on them obtain more comprehensive histories from patients (Suchman et al. 1997).

On the basis of this brief sketch, we may suspect that while clinical empathy entails a minimal form of personal understanding, it is best described as a *practical understanding* – as a skillful performance that unfolds based on embodied interaction with the patient.[5] Many of the features described under the banner "clinical empathy" are executed smoothly, without explicit awareness, knowledge-driven processes, or online performance monitoring. Such practical understanding is not about grasping explanations, but about certain abilities for embodied, engaged social interaction that require a second-personal, embodied stance and encompass the coordination of, for example, expressions, intonations, and gestures. In other words, it involves emotional and sensory-motor processes that are often described as "primary intersubjectivity" (Trevarthen 1979; Gallagher 2005).

Of course, this is not all there is to personal understanding in clinical encounters. In many cases, understanding illness will require understanding *reasons*. Those who favor the Davidsonian view that reasons are causes (Davidson 1963; for a discussion, see Risjord 2014, 88–91) could perhaps argue that personal understanding in such cases simply reduces to a sort of explanatory understanding, acquired, just like in scientific inquiry, by uncovering causal relationships of dependence. The only difference is that here understanding involves identifying relations of dependence between psychological elements that are causally involved in acting, thinking, or feeling. While this is not the right place to offer a thorough discussion of this complicated matter, in the ensuing examination of a particular type of situation involving *extended* forms of personal understanding, explanation and understanding appear to come apart. It is shown that extended personal understanding can be described as practical and may require adopting a particular second-personal stance and personal engagement.

It bears stressing that this distinction does not fully encapsulate the entirety of personal understanding. Some literature on social cognition highlights embodied ways to understand others that are constituted by skillful interaction (Gallagher 2005; Johnson 2015). Others distinguish between "understanding-about persons" and the more holistic "understanding persons" (Debes 2018). It should be pointed out that both minimal and extended personal understanding require more than "understanding-about persons" (which essentially boils down to knowing things

[5] Thanks to an *anonymous referee* for pointing out the importance of this aspect.

about persons), but less than "understanding persons" (which is too demanding in a clinical situation).

5.6.2 Extended Personal Understanding

Consider an interaction, loosely based on a published case study (see Snow and Fleming 2014), between a medical practitioner (MP) and a patient (P), a 74-year-old woman who complains of hoarseness, altered voice, and shortness of breath. After an initial medical interview, MP feels that she has obtained a clear enough picture. Based on her own observations and a battery of lab tests, MP has achieved some degree of biomedical understanding when MP has identified the variables upon which P's pathological condition depends: P has an upper respiratory tract infection and stridor and dysphonia caused by a large multinodular goiter, which compresses her trachea. MP grasps what produces the condition and its symptoms, allowing MP to make appropriate counterfactual inferences and accurate predictions about the course of disease and treatment. On this basis, MP leans toward recommending surgery (thyroidectomy).

At the same time, conforming to prevailing standards, MP also thinks that the right treatment must respect P's autonomy and reflect her values and goals in shaping her own future, particularly because the treatment might involve risks and affect the kind of life she will be able to lead. In this regard, current bioethics often emphasizes the importance of *respecting autonomy*, which is comprehended as the "acknowledgment of a person's right to hold views, make choices, and take action based on personal values and beliefs" (Beauchamp and Childress 2001, 61). To be able to suggest the optimal treatment, biomedical understanding of the pathological condition is not sufficient. Because MP strives to involve P as an active agent in her own care, she must present options in ways that are intelligible for P, and she must communicate medical knowledge in a way that enables P to consider the different options. Effective communication in such situations displays difficulties similar to those found in intercultural communication. As Laurence Kirmayer (2011, 413–4) puts it, due to the character of present-day medicine, there is a "cultural divide" between clinician and patient: "medicine constitutes a subculture with its own taken-for-granted background knowledge and, therefore, every clinical encounter is intercultural."

Bridging such a divide requires that MP is aware of broader aspects of P's history, background knowledge, and social environment, and a lack of personal understanding would hinder identifying a mutually acceptable treatment plan and MP's ability to care for P. To see why, consider how the interaction continued. MP recommends surgery to P and explains the risks, maintaining that the surgery is relatively safe, though she adds that one of the downsides is that it would leave a scar. Although P knows that she would most likely die from tracheal obstruction without surgery, to MP's surprise, she refuses treatment. P explains her decision by recourse to her desire not to live with a scar on her neck. In an important sense, MP has now gained knowledge of why P decided against the treatment option and is able to offer an explanation by citing the relevant belief-desire pair causally involved in P's decision. Nonetheless, while this offers MP knowledge that renders her able to offer an explanation of P's decision, it does not suffice for personal understanding in a richer, extended sense. Explanation and understanding come apart, and MP will likely report not being able to care for this patient because she is unable to understand her.

Given the tight link between explanation and understanding in theoretical understanding, the fact that they come apart in a clinical situation speaks in favor of interpreting personal understanding more in terms of practical understanding. MP's knowledge of the relevant belief-desire pair causally involved in P's decision is not enough for personal understanding. What is missing? Using a commonly accepted distinction between normative reasons (that justify an action as judged by an impartial observer), motivating reasons (that justify an action as judged by the agent), and explanatory reasons (see Alvarez 2016; Stueber 2017), we may say that the lack of understanding here boils down to MP's failing to comprehend the relevant belief and desire as constituting *a normative reason* for the decision. MP is unable to understand, because she assumes something about P's personal history, background knowledge, and larger network of defining values, including that the desire to prevent a worsening of the condition or death constitutes a much stronger reason than the aversion to a scar.

MP engages in further dialogue and realizes that she does not have any reason to assume that the rational capacities of the patient are impaired. The dialogue also reveals a crucial bit of information about P: in her native Sicily, bearing a scar on one's neck ("the Sicilian bowtie") references a violent Mafia practice and depicts the carrier as dishonorable. This information helps MP transition from being able to *explain* the patient's decision by comprehending the relevant explanatory or motivational reason, to being able to *understand* it as reflecting a basic value (being able to participate in social life) that MP can comprehend as being worth deeply caring about. Because this now resonates with values that MP cares about or at least comprehends as potentially worth caring about, MP is able to take up P's perspective and to immerse herself in it, for instance by using her own mind to reenact P's deliberative thought processes. Drawing on Stueber's (2017) account, a cognitively advanced form of *reenactive empathy* plays a key epistemic role in understanding reasons. On this view, one understands a person's reason if one is able to comprehend it as a reason that one could potentially entertain in the other person's situation. For our purposes here, we may leave aside the question whether such reenaction involves simulation as described in some of the literature on social cognition (see, e.g., Heal 2003; Goldman 2006).

This allows MP to grasp how the illness is woven into the fabric of P's life: what it means to P and how it affects the overall scheme of values, desires, and beliefs that are constitutive of who P is. MP is now able to comprehend how the illness experience and the envisioned therapeutic outcomes fit within a coherent narrative. P's decision is rendered intelligible for MP, and perhaps not entirely inappropriate from P's point of view.

In a sense, the extended personal understanding MP has achieved represents an epistemic gain that will allow her to optimally care for P. But perhaps this is nothing more than MP attaining personal understanding by grasping more explanations of P's motives, which could make us doubt that the relevant understanding here is practical. However, it is worth stressing that the epistemic gain occurs because MP is able to adopt a particular second-personal stance, an attitude characterized by *recognition*, that (counterfactually) understands the patient as a rational agent governed by normative reasons, with the decisions and actions of the person oriented toward ends that are worth pursuing.[6] In other words, understanding only becomes available to MP insofar as she adopts a particular stance toward the patient, as a being – just like her – whose life is centered on values that can be comprehended as being *worth* caring about.

5.7 Conclusion

Given that understanding is context sensitive and that medical research is fused with medical practice, it is important to clarify the specific nature of

[6] Drawing on Cavell's (1969) idea that our primary relation to others is not epistemic, Axel Honneth (2008) has argued that such attitude of recognition is a basic empathetic engagement that precedes cognitive access to other minds.

understanding in medicine and its particular contribution to supporting human agency. This chapter set out to make headway on the first issue, while Chapter 6 addresses the second one. After distinguishing types of understanding and connecting objectual understanding to grasping explanations, the chapter considered the history of scurvy to explore what obtaining understanding of a disease in the context of medicine involves. The main conclusion was that biomedical understanding of a disease requires grasping a mechanistic explanation of that disease. However, alluding to the distinction between understanding and explanation in debates on the methodological principles of the humanities and social sciences, it was argued that biomedical understanding is not sufficient for clinical understanding. Rather, clinical understanding combines biomedical understanding of a pathological condition with a personal understanding of an illness. In some cases, personal understanding is extended, necessitating the adoption of a particular second-personal stance and the use of cognitive resources in addition to those involved in biomedical understanding. Of course, personal understanding is a complicated matter, and some hold that it is neither theoretical nor practical. For example, Bengson (2017, 24) maintains that "a psychoanalyst's empathic understanding of a patient, or a lover's understanding of a beloved, is (perhaps) neither theoretical nor practical." The view proposed here describes personal understanding as practical, even if it acknowledges that it is not paradigmatically practical.

The conclusion has implications for what counts as progress in medicine, which will become clearer in Chapter 6, where we clarify the aim of medicine and its specific contribution to supporting human agency. We have started out by saying that while the *Understanding Thesis* and the *Systematicity Thesis* jointly imply that medicine has the character of a scientific inquiry that aims at understanding, it is natural to think that medicine aims to understand disease. It has become clear that this picture is not entirely accurate, as understanding illness could be shown to be a key component of clinical understanding. This begins to indicate that progress in understanding illness can count as progress in medicine, even without a simultaneous increase in biomedical understanding of disease.

CHAPTER 6

The Aim of Medicine I
The Autonomy Thesis

6.1 Introduction

In contemporary debates on the future of medicine, it is customary to interpret pressing challenges to medicine in the twenty-first century as calling for reforms in the manner in which medicine is financed, organized, managed, and delivered. Consequently, much attention is devoted to issues like the role of the market in meeting the growing needs of an aging population, controlling costs by centralization and monetary incentives, and the value of new digital technologies (e.g., artificial intelligence, 3D printing, and robotics) in supporting personalized medicine. While these are obviously highly important matters, the focus on financial and organizational matters seems to assume that the aim of medicine is obvious and well established and the task is to devise the most suitable and effective approach to its realization.

Such an assumption, however, is mistaken. The aim of medicine is far from obvious and well established, and in light of the challenges to medicine discussed in Chapter 1, as well as other factors (e.g., the increasing economic pressures on medicine, and the social, moral, and political implications of the expansion of medical knowledge), attaining clarity on this matter is more pressing than ever. To assist progress, the main task of this chapter is to evaluate several options and propose an account of the aim of medicine. While some have attempted to catalogue a number of goals that medicine permissibly pursues (see, e.g., Callahan et al. 1996; Brody and Miller 1998; Boorse 2016; Schramme 2017a), the chapter will involve a critical interchange with an attempt by Alex Broadbent (2019) to identify a single, overarching goal.

Findings of earlier chapters offer important guiding impulses for completing this task. The *Systematicity Thesis* and the *Understanding Thesis* led to the broad suggestion that the aim of medicine is to understand pathological conditions in order to contribute to the endeavor of supporting human agency. While aspects of understanding in medicine were investigated in Chapter 5, we now turn our attention to clarifying in more detail *how* medicine contributes to supporting human agency. For this, the chapter examines the opening proposal according to which medicine is pathocentric, aiming to promote health by curing disease. Discussing and rejecting this opening proposal as well as competing ideas, the chapter presents and defends the *Autonomy Thesis*, which holds that medicine is not pathocentric, but aims to promote health with the final aim to enhance human autonomy. The chapter adopts a "positive" notion of health, clarifies its relations to other concepts such as well-being and autonomy, and offers a pluralist perspective on some difficulties surrounding the concept. It closes by considering and defusing the objection that the *Autonomy Thesis* is overly permissive and allows many highly controversial procedures as legitimate parts of medicine.

The talk of *the* aim of medicine is bound to raise the eyebrows of some readers. Can a multifaceted enterprise that is a diverse subject across the globe be encompassed by a single aim? It is therefore important to add two specifications. First, the claim is not that every single action performed in the context of medicine is directed at this aim or that it is something that all participants of medicine share. But that said, even complex institutions are held together only if most of their activities share a methodical effort to achieve some *constitutive aim* (of a limited set of constitutive aims) that they could not fail to pursue without losing their identity, and that governs what counts as progress in that activity (see, e.g., Bird 2019a). Second, our inquiry is limited to "mainstream medicine" (i.e., scientific Western medicine) that – at least on some level of abstraction – is sufficiently universal in spite of variation in the cultural meaning of crucial concepts (i.e., disease, health, sickness, illness, pain, and disability) in local features of institutions and practices, and in its societal role across societies and cultures (for a discussion, see Broadbent 2019, ch. 1).[1]

Keeping such limitations in mind, it is plausible to speak of mainstream medicine as a universal discipline in a broad sense: it serves basic human needs in dealing with health and disease that are sufficiently shared throughout humanity, and it does so while adhering to central values (e.g., informed consent, the benefit of the patient, grounding treatments on the best available scientific evidence) that in part constitute medicine's identity. On some level of abstraction from local features, medicine is a

[1] Due to this focus, the view offered here has no direct implications for thinking about veterinary medicine.

coherent enough enterprise to have a universal aim. It is this aim that patients as well as medical professionals sometimes make recourse to when they criticize "local" variations in medical practice or when they complain that medicine is under sociopolitical pressures to serve "alien" purposes.

Two methodological considerations guide the chapter. First, the question about the aim of medicine is approximated by deploying a sequential approach that proceeds by articulating increasingly more nuanced theses. The chapter starts with common assumptions, identifies a problem, and suggests a more complex iteration, which is introduced in response to the problem. Instead of simply describing and defending the final iteration, using such a sequential approach helps tease out more or less articulated ideas, which put us in a better position to identify different conceptions of health and disease and to clarify their relations to each other and to other concepts such as well-being and autonomy. Outlining an intelligible route in which the steps build appropriately on their antecedents will make it easier to achieve reflective equilibrium.

Second, the question about the aim of medicine is explored in tandem with a closely connected matter that concerns the *internal morality of medicine*, that is, the moral norms and values that govern the practice of medicine (for recent discussions, see, e.g., Symons 2019; Hershenov 2020; Ng and Saad 2021). To comprehend what the internal morality amounts to and how it relates to the aim of medicine, it is helpful to call into mind that clinical medicine is a *practice*, a social activity that has a teleological structure, defined by particular goods.[2] For example, education is a practice that is aimed at developing the rational and affective capacities of human beings, and this gives rise to a set of professional norms that determines how "excellence" is understood within the framework of education. Clinical medicine displays a similar teleological structure with the aim, most commonly assumed, to combat pathology and enable patients to engage in activities and to increase their quality of life. Excellence in the context of clinical medicine is tied to providing effective treatment options for patients, and participants in the practice are subject to norms that jointly constitute an "internal morality" of the practice. These norms generate prima facie moral obligations on medical

[2] On Alasdair MacIntyre's definition, practice is "any coherent and complex form of socially established cooperative human activity through which goods internal to that form of activity are realized in the course of trying to achieve those standards of excellence which are appropriate to, and partially definitive of, that form of activity" (2007, 187).

professionals independently of general morality and offer a normative backdrop against which inappropriate use of medical understanding can be identified.[3]

Importantly, this internal morality cannot be reduced to general morality (i.e., the collection of moral norms in a given society) because the two can be at odds with each other. First, the internal morality can *suspend the restrictions* of general morality. Consider, for example, confidentiality. If someone discloses to us information in confidence, general morality dictates that we maintain confidentiality, but allows a number of reasons for breaking confidentiality. In clinical medicine, even when general morality would dictate breaking confidentiality, the internal morality of medicine might suspend this norm, permitting physicians to maintain confidentiality.[4] Second, the norms of internal morality can *add to the restrictions* of general morality. While the latter may allow participation in certain activities (e.g., torture, execution, forced sterilization, using pharmaceuticals to render prisoners passive, use of human subjects for research without informed consent), there is widespread agreement that the participation of medical doctors in these practices would represent a violation of the internal morality of the profession. The norms render certain acts immoral for physicians that might otherwise be morally permissible and vice versa, restricting the proper use of medical means to the pursuit of the goal of the practice.[5]

The idea of an internal morality has attained a prominent place in the philosophy of medicine and medical ethics. While there is significant disagreement in literature on a number of its aspects (e.g., to what extent it is fixed, to what extent it is autonomous from general morality), what is common to the different accounts is that they trace the moral norms and values that govern the practice of medicine and aim to distinguish legitimate practices from those that violate the internal morality of medicine.[6]

[3] Importantly, these norms can be read off by studying the practice itself from the outside, and are thus not only available from the vantage point of health care professionals (Veatch 2001).

[4] Also, confidentiality acquired in the course of practicing medicine does not require any additional act of promising.

[5] Acts that run counter to these goals are prohibited, but can in some cases be overruled by external morality.

[6] For example, proponents of "essentialism" (e.g., Pellegrino) hold that the telos and the good at which medicine aims (the health of the patient) are fixed and so are the duties that they generate. Proponents of "evolutionism" (e.g., Miller and Brody 2001) hold that the aim of medicine can be subject to change along with the duties that it gives rise to (see Ng and Saad 2021).

6.2 Cure and Treatment

We start with the commonly held view according to which medicine is pathocentric, in the sense that it aims to promote health by curing disease. To grasp what this view involves, we need to add some clarification about what "health," "disease," and "cure" mean in this context. On a standard account, health is the absence of disease, where disease encompasses not only what are usually considered as prototypical diseases (e.g., infectious and chronic diseases), but also conditions that are roughly comprehended as deviations from some range of normal functioning (e.g., injuries, poisonings, growth disorders, functional impairments) (see, e.g., Boorse 2014). Cure is usually comprehended as an intervention that leads to the full elimination of the disease, but does not require that the patient returns to a state that she would have been in had she not been afflicted by the disease. Clearly, curative interventions may count as successful even if psychological or bodily injuries suffered from a disease remained unaddressed, and even if the marks of the intervention have a lasting impact on the patient's life. For example, a successful curative intervention by surgery may be accompanied by the formation of scar tissue, curing an infection with antibiotics may cause irritation of the stomach lining, and curing affective disorders with antidepressants may lead to weight gain.

Having added these clarifications, we may now return to the view that medicine is pathocentric, in the sense that it aims to promote health by curing disease. An obvious objection to this initial proposal is that there are a large number of medical interventions that do not offer a cure. And yet, if somebody consults a physician with a bad case of the flu, hepatitis B infection, or asthma, some kind of intervention of a properly medical nature will take place. But the purpose of the intervention in such cases cannot be a cure, because there is none. Instead, the purpose is *treatment*. In the medical literature, the relationship between cure and treatment is not always clear. One encounters both sentences like "there is no true cure for *x*, but it can be treated" (which implies a distinction between cure and treatment) and sentences like "minor infection *x* can be treated" (which seem to collapse the distinction between cure and treatment). However, in general, "treatment" encompasses both the *cure* and the *management* of conditions.[7]

[7] Treatment will often include a range of activities that in themselves do not amount to treatment such as the disinfection of the skin prior to injections or mild sedation prior to an uncomfortable intervention.

A treatment can amount to a cure if it eliminates the disease causing the symptoms, like antifungal ointments curing athlete's foot by killing the fungus that causes it. While cure requires eliminating the disease, many treatments like insulin injections for diabetes suppress or reduce symptoms and mitigate harm caused by disease. To take another example, there is no cure for chronic hepatitis B infection, but there are a number of approved medical treatment options that can effectively manage the disease, by controlling the virus and reducing the risk of developing serious liver disease. Similarly, there is no cure for asthma, but there are ways to treat it to reduce the likelihood of an asthmatic episode. In some cases, treatment requires that medication is administered consistently for the patient's entire life. Such management might keep the patient completely symptom-free for her entire life, but it does not amount to a cure as it does not remove the underlying cause.

The distinction between cure and treatment notwithstanding, one could insist that the final aim of medicine is still cure, while treatment – or clinical management that reduces symptoms or alleviates suffering or harm caused by disease – is what clinical medicine does if cure is not possible.[8] But this is not entirely correct. While a detailed argument would lead us off path, it is perhaps enough to call attention to commonly occurring cases in which there is no cure, but a treatment is offered that effectively reduces the risk of developing sequelae (other diseases such as pneumonia in the case of the flu, or other liver disease in the case of hepatitis B). However moderate, such a treatment is still regarded as successful and as progress toward the aim of medicine, even though no pathological condition has been cured. If the aim were cure *simpliciter* then we would reach a counterintuitive result: treating a condition to thereby prevent other diseases from emerging would not count as making an advance toward the aim of medicine.

6.2.1 *Treatment and Care*

Treatment thus encompasses a range of activities and includes cure and management. There is, of course, more to say about fine-grained distinctions between cure, treatment, and management, but for our purposes the previous considerations provide enough footing to take a step toward addressing some of the problems mentioned in Section 6.1. Avoiding the

[8] A similar general view is defended in Broadbent (2019). A detailed treatment of Broadbent's position will ensue in Chapter 7.

challenges with "cure" outlined earlier, we may adopt as our working proposal that the primary aim of medical understanding is to promote health by treating disease. Still, one might argue that the example of prevention, especially with respect to diseases that are very unlikely to ever occur, spells trouble for this proposal, because describing prevention as medical treatment appears forced. The trouble is that while prevention cannot be readily incorporated into the concept of treatment, there is still a sense in which it is an instance of medicine being practiced, and successful prevention counts as making progress toward the aim of medicine. To resolve this issue, we start by exploring two possibilities.

The first possibility is to argue that because such preventive interventions occur before diseases are actually manifest, prevention is not a part of medicine like treatment is. It merely *uses* medical understanding, in a somewhat similar fashion as medical understanding can be used to construct ergonomic computer keyboards or air-conditioning systems that minimize the risk of transmitting infectious diseases in a building. However, this possibility fails to withstand scrutiny, because vaccination can in most cases be seen as a form of risk management, which is structurally isomorphic with interventions that are conventionally regarded as a part of medicine (e.g., vaccination, the treatment of risk conditions like hypertension, surgical removal of precancerous tissue). Maintaining that prevention is not a genuine part of medicine like treatment is would thus result in a highly implausible view.

The second and more appealing possibility is to deny that prevention is an instance of medical understanding serving some nonmedical aim, but – acknowledging the difficulties with incorporating prevention into the concept of treatment – to replace "treatment" with "care." This allows us to effortlessly incorporate prevention along with other cases that we more readily describe as treatment (including cure and management). In this manner, we can think of medical care as including treatment and prevention as subordinate goals to the same end, namely, eliminating disease or rendering its occurrence much less probable. If treatment and prevention are part of a larger goal of promoting health, then we may propose that the aim of medicine is to offer care that promotes health by treating and preventing disease.

This line of reasoning has perhaps convinced our interlocutor that prevention can be accommodated into our current proposal, but she can still point to difficulties in accounting for common medical interventions like pain relief, which by all accounts constitutes a central medical activity. Although the administration of pain medication might be delayed for diagnostic reasons or to minimize the risk of addiction, it is unambiguous that medicine places significant emphasis on alleviating pain. Yet, our interlocutor might argue that pain relief does not fit the current proposal, because it simply does not promote health by treating and preventing disease. Instead, pain relief exemplifies an important *use* of medical understanding to pursue another aim, but not an aim *of* medicine, even if it is performed by medical professionals using medical techniques in regular hospitals or medical facilities. Alex Broadbent (2019, 51) has proposed a similar view, stressing, for example, that palliative care only enters the picture when cure and treatment had been given up. As he puts it, "palliative care is consistent with my assertion that pain relief is a use of medical skills and tools, but not a goal of medicine" (Broadbent 2019, 50).

In reply to such a position, we have at least two strategies at our disposal. The first starts by distinguishing between pain as a symptom of injury or disease and chronic pain. The latter may be considered as a disease in its own right, because it has developed into a destructive force that no longer has a beneficial function (Raffaeli and Arnaudo 2017).[9] If we accept that chronic and recurrent pain is a disease, then palliative care is still covered by our proposal, as it promotes health by confronting disease. Nonchronic pain is, of course, a different matter, but perhaps relieving nonchronic pain could be understood as a preventive measure, as pain leads to mental and bodily stress reactions (e.g., increase in blood pressure and heart rate) that weaken the immune system and increase the risk for conditions such as heart disease. If these considerations are on the right track, then the relief of both chronic and nonchronic pain would still fit the proposed account that the aim of medicine is to provide care, which promotes health by treating and preventing disease.

The second strategy contends that even if pain could not be considered as a disease in its own right, it is not clear that palliative care is a use of medical understanding for nonmedical purposes. To see why, consider cases of terminal conditions where palliative care is pursued not in tandem with potentially life-extending treatment, but, after weighing the risks and benefits of treatment, instead of it. These are typically cases in which treatment options offer little chance of extending life, and the prospect of returning home and staying pain-free for the remaining time seems more attractive than the prospect of gaining some time but suffering the side

[9] This is why the European Federation of IASP Chapters' declaration on pain maintains that "chronic and recurrent pain is a specific health care problem, a disease in its own right" (Raffaeli and Arnaudo 2017).

effects (e.g., post-surgery pain, chemotherapy-induced nausea). An implication of the position that Broadbent proposes is that in such cases, if physicians advise palliative care over the attempt to pursue life-extension, they would be advising a course of action that entails using medical understanding to pursue a nonmedical aim *instead of* pursuing the aim of medicine. But in that case, we run into a problem. While the course of action the physicians advise is permissible and relatively standard, it counts on Broadbent's view as the pursuit of a nonmedical aim over the pursuit of the aim of medicine. The problem is that such acts would normally be considered inconsistent with, and indeed as violating, the internal morality of medicine.

Importantly, these complications dissipate if we adopt the suggestion to use "care" as describing the aim. Doing so not only offers a resolution to the problem with accommodating palliative care; it also allows us to comprehend palliative care not merely as a nonmedical aim that physicians can legitimately pursue, but as something that actively promotes the aim of medicine.

6.3 Medicine Is Sanocentric, but Not Necessarily Pathocentric

While accommodating palliative care helps clarify the idea that the aim of medicine is to deploy medical understanding to care for individuals and to promote health, further reflection on its nature raises reservations about the idea that care necessarily occurs by treating and preventing disease. If it is true that pain constitutes something like a disease in its own right, then we may maintain that palliative treatment is consistent with the idea that the aim of medicine is to care for individuals and to promote health *by treating and preventing disease.* But whether or not this argument is successful, there are many other cases of what appear to be genuinely medical activities that do not tackle disease, but are still hard to comprehend as pursuing nonmedical goals. Consider the following two examples:

> *Age-related sarcopenia.* A patient in her eighties consults her physician with concerns about loss of muscle strength. Having excluded potential underlying diseases, the physician explains to her that her condition is known as sarcopenia, which is considered not to be a disease, but a common condition associated with old age, caused by a process that gradually reduces muscle tissue and increasingly replaces muscle fibers with fat tissue. The physician informs her that the most effective method to moderate this process is regular strength training combined with protein-enriched nutrition.

> *Pregnancy nausea.* A 32-year-old woman in the first trimester of her pregnancy consults her physician with nausea and occasional vomiting. Having excluded potential diseases (e.g., hyperemesis gravidarum), the physician explains that especially during the first trimester of a pregnancy, nausea and occasional vomiting are perfectly normal and impact approximately two-thirds of pregnant women. She advises eating small meals and limiting spicy and acidic foods, and prescribes an antihistamine (promethazine) in case the condition does not improve.

In each of these cases, the consultation involves two phases. In the first phase, biomedical understanding is deployed to specific or nonspecific symptoms in order to assess and exclude the possibility of underlying disease. In this phase, there is little doubt that medical understanding is being used for genuinely medical purposes: the likelihood of a number of diseases based on a set of symptoms is evaluated, which is a form of risk reduction that belongs to the same category as prevention.

In the second phase of the consultation, one might argue that something else ensues. Disease drops out of the picture as age-related sarcopenia and pregnancy nausea qualify neither as diseases nor as genuine risk factors. Instead, the activity of the physician during the second phase of the consultation seems to pursue nonmedical aims, perhaps best described as furthering the well-being of the patient in some broad sense. If we were to hold on to our previous thesis (i.e., the aim of medicine is to care for individuals and to promote health by treating and preventing disease), then we would be forced to accept that medical understanding is used for nonmedical purposes. The relevant premises and conclusions can be laid out as follows:

(P_1) The aim of medicine is to promote health by treating and preventing disease

(P_2) Health is the absence of disease

(P_3) Age-related sarcopenia and pregnancy nausea do not involve disease

(C_1) Thus, in the second phase of the encounter, the physician's use of medical understanding does not promote health

(C_2) Then, in the second phase of the encounter, the physician uses medical understanding for nonmedical purposes

While the argument is valid, the conclusions put us into an awkward position for at least two reasons. First, there are many similar cases in which medical attention is directed at typical conditions of old age (e.g., frequent urge to urinate due to reduced bladder capacity, weak hand grip, thinning of the epidermis and dermis, xerosis, refractive errors, back and neck pain, feelings of exhaustion) that are not considered diseases. These are so numerous that accepting (C_2) would force us to accept the somewhat strange consequence that a significant percentage of medical interventions do not pursue the aim of medicine.

Second, a weightier reason for being alarmed about the conclusion is that accepting (C_2) would lead to a clash with intuitive judgments about some of the norms in medicine that make up part of its internal morality. Usually, this internal morality serves as a normative backdrop against which inappropriate uses of medical skills and knowledge can be identified. Here, we may attempt a different strategy and start by noting that under normal circumstances, as long as a physician operates within her area of expertise and has the required resources, the internal morality only permits limited leeway to refuse to provide treatment to a patient that she has formed a therapeutic relationship with. There are cases, however, in which physicians can refuse without penalty interventions that do not serve medical aims.[10] In such cases, the physician can legitimately opt out by appealing to the nature of the profession even if refusing the requested intervention might violate external moral (or perhaps legal) norms. Such cases are unlike those in which internal morality forbids participation (e.g., torture), and more like cases in which refusal can legitimately ensue on grounds such as medical futility or patient noncompliance impeding the physician's ability to provide proper care.

The important point is thus that in cases in which no medical aim is involved, the internal morality of medicine does not place the physician under obligation to help. Put in slightly more precise terms, a reflection on the internal morality of medicine lets us conclude that

> (T) there is no professional obligation to use medical understanding for interventions that serve nonmedical aims.

This line of reasoning is not restricted to medicine. Although it does not apply for all types of professional activity, for certain types we can say that if X is the goal of a professional activity or practice that one has "signed up for," then one may opt out of doing something with the goal Y, even if Y is not inconsistent with X. For example, consider a teacher who has formally adopted the profession, the aim of which is presumably something close to expanding the cognitive capacities of school children such that they can become competent citizens. Imagine that he is asked by the school administration to transport children to school, which is not forbidden by

[10] For a helpful discussion of such cases in reply to Boorse's account, see Hershenov (2020).

the internal morality of the profession and is not inconsistent with expanding the cognitive capacities of school children. But while not prohibited (although it might require an additional safety clearance), it is not mandatory either. The teacher could choose to help without violating the internal morality of the profession, but he could also stress that it is not required by the norms of the profession, and he could decline the request by saying "this is not what I have signed up for" without deserving reproach for violating the internal morality of the profession. Moreover, he could also stress that fulfilling the request would actually take time away from teaching and mentoring activities that constitute professional duties. In any case, upon declining, the administration could perhaps complain that this teacher is violating external morality (i.e., he refuses to "take one for the team"), but they would not be justified in blaming the teacher for violating professional norms.

We may now revisit (C_2) in light of (T). Accepting (C_2) means accepting that in the cases of age-related sarcopenia and pregnancy nausea, the physician uses medical understanding for the benefit of the patient, but does so while pursuing a nonmedical aim. This, however, has significant consequences if we simultaneously accept (T): helping to alleviate the discomfort caused by nausea and providing information about the most effective approach to moderate sarcopenia are now neither prohibited nor mandatory. This means that the physician could choose to help without violating the internal morality of medicine, but she could also refuse treatment without violating the internal morality of medicine. Alleviating the discomfort and informing the patient would then be something like a charity-based or *supererogatory* action, which is optional, beyond any professional duty, and the omission of which would not deserve criticism.

Consequently, if we accept (C_2) and (T), we obtain a result that is hard to accept: by appealing to the norms of the profession, the physician could have refused without penalty to provide information about the most effective approach to moderate the loss of muscle tissue or to treat the nausea. However, when assessing such a course of action, the expected reaction would be that there is something amiss with this physician's comprehension of the norms of the profession. But if we do not accept (C_2), then we have to reject at least one of the premises. But which one(s)?

6.3.1 Sanocentricity and Positive Health

The suggestion here is to accept (P_3), modify (P_1), and replace (P_2). (P_3) stands firm because cases like *age-related sarcopenia* and *pregnancy nausea* do not involve disease by either lay or professional standards. However, holding on to the view that medicine is sanocentric, we may modify the first two theses as follows:

(P_1*) The aim of medicine is to promote health.

(P_2*) Health is more than the absence of disease (positive health).

Accepting these theses would allow us to acknowledge that, in the cases under consideration, the activity of the physician during the second phase of the consultation can be adequately described as using medical understanding to further the aim of medicine, namely the promotion of health of the patients by increasing their robustness or resilience. Medicine would then qualify as sanocentric on some broad, positive sense of "health," and we could maintain that while medicine often proceeds by treating and preventing disease, it is not restricted to it. But are the modifications introduced acceptable?

The modification in (P_1*) is relatively uncontroversial, as it only introduces changes as to how the promotion of health is achieved. It is also a better fit with the treatment of pain and smaller injuries in cases that cannot be readily apprehended as involving disease or preventive measures. The second modification might be more difficult to agree to. Some might think that accepting (P_2*), thus adopting a positive notion of health and giving up a negative notion on which health is merely the absence of disease, is implausible: it is quite natural to say that when a person has a disease then she is not healthy, and when a person is healthy then she does not have a disease. But there are at least three reasons for thinking that the notion of health is not entirely coupled to the absence or presence of disease.

The first reason is that health as the absence of disease is not consistent with how the general enterprise of health enhancement is understood in modern societies. Following Lennart Nordenfelt (1998), we may distinguish between two types of health-enhancement activities in modern societies: *health care* (medical care, nursing, rehabilitation, and social care) and *health promotion* (health education, medical prevention, environmental care, and legal health protection).[11] *Health care* intends to

[11] The distinction appears somewhat artificial, because under normal circumstances, caring about *x* involves promoting *x*. Thus, health care seems to entail promoting health. Another issue is that medical prevention only figures in the category of health promotion. Of course, both clinical medicine and public health measures aim to preserve and promote health by various interventions. But there is an important distinction between preventive care in a clinical setting and the kind of preventive measures that offer health education via distributing flyers or TV advertisements.

improve or support the recipient's health, which is perceived as in danger due to the presence of injury, disease, or some other internal risk factor. Medical care aims at eliminating diseases, injuries, and risk factors and reducing their negative consequences, while nursing aims to care for the ill by supporting vital functions during a period of illness. Rehabilitation (i.e., interventions designed to optimize functioning) and certain forms of social care are intended to support health typically after the individual has been stabilized. *Health promotion* can be directed to improve or support the health of individuals, groups, and entire populations without presupposing some reduced state of health.[12] For instance, health education and prevention efforts are familiar, while environmental care and legal health protection typically involve regulating substances perceived as potential health risks (e.g., alcohol, cigarettes, pollution, and stress in work environments).

This outline of the general enterprise of health-enhancement supports the picture that prevention through inoculations, screenings, preventive surgery, and medical education (e.g., in the form of physicians advising preventive lifestyle changes) is not medical understanding being used for nonmedical aims, but a part of medicine on a par with treating already existing conditions (e.g., diseases, injuries, deficiencies). However, it also shows that health-enhancement activities in modern societies operate with a positive notion of health (i.e., health is more than the absence of disease). For example, if rehabilitation efforts designed to optimize functioning in everyday life after the pathological condition is cured count as *health* care, then health must be more than the absence of disease. Of course, this is not a knock-down argument against comprehending health as the absence of disease, but it does offer a first indication.

The second reason for thinking that (P_2*) is correct is that using a conception of health on which health is simply the absence of disease has problematic implications. If health is merely the absence of disease, then it is difficult to make sense of comparisons between degrees of health that both lay people and medical professionals regularly make. For example, it is possible to say of two individuals with disease that one is unhealthier than the other, and we can also compare the health of people

[12] Nordenfelt maintains that the reason for the health-promoting act is not the recipient's state of health. But that seems to overlook interventions on certain segments of the population that may be initiated out of a concern for their low degree of initial health state, without rendering them an actor of health care.

across generations, maintaining, for instance, that despite longer life expectancy, presently living adults are actually less healthy than were adults in previous generations. Some think that such comparisons indicate that "health" may be a fundamentally comparative concept (Schroeder 2013), but the important point here is a different one: if health were merely the absence of disease, then we would lack adequate resources to compare two healthy people. We can say that A and B are healthy because they do not have a disease, but we cannot say that A is healthier than B, which seems insufficient, as being healthy can involve a spectrum of states of healthiness. So if we deem A healthier than B, then we are attributing to A something that B has less of, but whatever it is, it cannot be disease. Section 6.4 looks more closely at puzzles generated by comparisons of this kind. In reply one could say that such health comparisons do not necessitate positive health, but only some standard of being in perfect health, characterized by a complete absence of disease. Whether one is more or less healthy would then simply depend on one's distance from that standard. However, if being in perfect health is characterized as a complete absence of disease, then relatively large groups of individuals would qualify (e.g., a sizable part of first-year university students would probably qualify). But then, we would encounter problems with explaining on what basis we compare the health of individuals in these groups.

The third reason is that debates on chronic disease and disabilities also seem to indicate a conception of health that is more than the absence of disease or disability. It is now increasingly common to think that it is possible to be healthy while having impairments that count as disabilities. As the Centers for Disease Control and Prevention (CDC 2020) clearly states, "Having a disability does not mean a person is not healthy." Moreover, it is increasingly accepted that it is unsatisfactory to indefinitely classify persons with a successfully managed common chronic disease as unhealthy (Ventakapuram 2013). But, even if the condition can only be successfully managed, neither lay people nor medical professionals would claim that it excludes being healthy.

These reasons offer support for (P_2*), and thus the thesis that health is more than the lack of disease. But, if this is true, then it opens up the possibility for thinking that medicine is not limited to promoting health (in the sense of the absence of disease). With respect to cases like age-related sarcopenia and pregnancy nausea, (P_2*) would allow understanding the second phase of the respective clinical encounters as aiming to promote health although without confronting disease. For this, however, we need to say more about health beyond adopting a positive notion and hence acknowledging that it is more than the lack of disease. This task will be pursued by offering a *pluralist perspective* on some puzzles and difficulties surrounding the concept of health.

6.4 Health: Two Puzzles

We start by taking a closer look at what appear to be two puzzles about health. The first puzzle is that the analysis of health generates very different intuitions (for a discussion, see Kingma 2019). Some think that the concept of health as the absence of disease is clearly intuitive while others think that it is clearly not. While "surface" intuitions of lay people lacking philosophical training can often yield unclear results (Kauppinen 2007), one might suspect a prima facie problem if rigorously trained professional philosophers eliciting properly directed intuitions about a circumscribed subject fail to reach at least some consensus (Sosa 2007). Such conflicts should make us cautious in our applications of the relevant concepts (see Williamson 2004) and should also motivate us to determine its source and possible implications for our inquiry. Clashing intuitions might sometimes indicate that the intuitions elicited are themselves systematically biased in some sense (Schroer and Schroer 2013), or that there is something irregular with the concept under analysis.

The second puzzle arises upon examining "healthy" as a gradable and comparative adjective (see Schroeder 2013). Adjectives fall into two subclasses, gradable and nongradable, where *gradability* is a semantic property that identifies different degrees of the quality it denotes. Gradable adjectives (a) admit comparative (taller) and superlative forms (tallest) and (b) can be modified by an intensifying adverb (e.g., *fairly, rather, less, very*). "Healthy" is a gradable adjective that (a) admits both comparative and superlative forms (healthier, healthiest) and can be modified (less healthy, very healthy, etc.), whereas none of this is true for nongradable adjectives (e.g., *wooden*). Not all gradable adjectives are gradable in the same way, but if we hold onto the idea that health is merely the absence of disease, then using comparative forms generates contradictions. Consider the following comparison in health on the level of the whole person.

(a) A is unhealthy. He has celiac disease, an immune reaction to eating gluten, which can damage the intestinal lining, prevent absorption of some nutrients and cause diarrhea, fatigue, weight loss, and anemia. Fortunately, A never eats gluten.

(b) B is healthy. Many of B's vitals (e.g., blood pressure, cholesterol, triglycerides, BMI) are, however, very close to being abnormal.

(c) A is healthier than B.[13]

Consider also a similar comparison on the level of single organs.

(a′) A is the unhealthy kidney of a 20-year-old. It is subject to a fully controlled kidney disease that reduces its functioning by 15 percent.

(b′) B is the healthy kidney of a 92-year-old.

(c′) A is healthier than B.

While (a) and (b) follow from the view that health is the absence of disease, (c) is intuitively appealing. And yet, together, they generate a contradiction. How could somebody who is healthy turn out to be less healthy than someone who is unhealthy? It is exceedingly counterintuitive that somebody who lacks health due to the presence of a disease should be more healthy than someone who is healthy. In the same way, when comparing single organs, (a′) and (b′) follow from the view that health is the absence of disease, yet they clearly collide with (c′). If both organs are up for donation and one has to choose the healthiest one, it is very likely that the choice will fall on A. Also here, counterintuitively, an unhealthy organ could turn out to be healthier than a healthy organ.

In light of the two puzzles surrounding "health," we may suspect that they are generated at least in part because there is something amiss with the concept itself. In that case, "health" would be less illuminating in explanations and generalizations, which would make it an appropriate target for conceptual engineering and give us reasons to pursue epistemic revisionism in order to improve the defective concept. To make progress here, we proceed by taking a closer look at the concept of health, but deviate from the standard analysis and engage in conceptual engineering as introduced in Chapter 2. The guiding idea is that the puzzles indicate that "health" admits of multiple incompatible analyses and that interlocutors might take "health" to express closely related, but different properties. The suggestion here is thus not that all we need to do is to equate health with positive health. Instead, the suggestion is to endorse a form of *conceptual pluralism*,

[13] The example draws on structurally similar cases from the work of Schroeder (2013) that deal with intergenerational assessments of health. Schroeder argues that while theories of health and health metrics are exclusively noncomparative, "health" may be a fundamentally comparative concept: "healthier than" is a conceptually more fundamental judgment than "healthy." The aim here is a different one, limited to a discussion about a negative vs. positive concept of health.

which acknowledges that different things deserve to be called "health" and holds that the puzzle is most likely generated by the activation of two slightly different concepts of "health." The task of conceptual engineering is to identify them, to articulate the different roles they play, and to reflect which ones are best suited to play what roles in our inquiry.

The idea is that "health" is stretched out between two concepts, and the modifiers "negative" and "positive" will be used to help express the concepts while allowing a degree of continuity. One important difference is that negative health refers to a *state* characterized by the absence of disease, and positive health refers to the presence of a *capacity*. The latter can be conceptualized in a number of different ways, for example, as some resilience or robustness of organisms or systems dealing with stressors and internal disturbances, or as Lennart Nordenfelt (1987; 2007; 2017; see also Venkatapuram 2013) has suggested, as the second-order ability to achieve vital goals. While such a broad characterization is satisfactory for our purposes, the capacities or abilities linked to positive health can be rendered more precise in a number of ways. For example, resilience and robustness can in general terms be comprehended as linked to maintaining set ranges of functioning (e.g., homeostasis) at low cost, or being able to rapidly return to a previous functional level following perturbation (Ananth 2008; Sholl and Rattan 2020). While a detailed treatment of this issue is not our current focus, it seems beneficial to separate robustness (ability to resist deviation from the original state) from resilience (recover after a deviation from the original state) for a more thorough theory of health (Ukraintseva, Yashin, and Arbeev 2016).

We should add that positive accounts in the literature are often left vague, and some defenders of a negative concept of health like Boorse (1975; 1977; 1997; 2014) will be critical of the prospects for a useful notion of positive health. For example, Boorse links positive health to some level of functioning that is above normal, and maintains that positive health conveys "the vain hope of giving personal or social values the objectivity of traditional medicine" (Boorse 1977, 572). However, it is not clear why positive health as the presence of some *capacity* cannot be conceptualized in a way that is immune to this problem. One could conceptualize positive health in terms of the ability of the organism to live through a range of likely future environments (Kingma 2012) or the second-order ability to achieve vital goals (Nordenfelt 1987; 2007; 2017; Venkatapuram 2013), or by deploying newer concepts (e.g., allostasis, resilience, plasticity, robustness, homeodynamics) that the literature in physiology and pathophysiology often takes to track specific properties of healthy functioning in a positive sense (Sholl 2021). Why should we think that conceptualizing positive health in all these ways will be based on "personal or social values"? A thorough discussion of this matter, as well as providing an account of positive health, is beyond the aims of this chapter. The focus here is on the *Autonomy Thesis*, which, as I noted, is compatible with all these accounts of positive health. However, due to its distinction between well-being and positive health, it is not compatible with accounts that identify positive health with well-being (e.g., WHO).

With this in mind, let us return to our puzzles. The first puzzle is relatively easy to deal with if we accept that "health" is stretched out between two concepts. As already indicated, the analysis of "health" generates clashing intuitions because in some cases it activates the negative concept whereas in others it activates the positive concept. The second puzzle can be dissolved along similar lines. Returning to our comparison of A and B with respect to their health, it is uncontroversial that the negative concept of health is active in (a) and perhaps also (b). However, when we get to (c), there is a shift to positive health, which allows that in spite of A's celiac disease, A may well be much more resilient or robust when dealing with stressors and internal disturbances. In a similar fashion, negative health is active in (a′) and most likely also (b′), but we shift to positive health in (c′), on which A may still be healthier than B given the aggregate costs of healthy functioning and successfully adapting to disturbances throughout a long lifespan.

Overall, unlike negative health, positive health does not exclude deficiencies due to well-controlled diseases, and this is true for both entire organisms and single organs. Positive health is not necessarily inconsistent with disease or disability, because the relevant capacities are not necessarily significantly reduced by the presence of disease.

6.5 Positive Health and Well-Being

Before we return to discussing the aim of medicine with the help of the positive notion of health, we should add that while positive health is closely associated with and causally linked to well-being, the account proposed in this chapter does not support stronger views according to which health simply is a kind of well-being.

Many approaches to well-being hold that health is not an end in itself, but is instrumental to and at least partially derives its value from larger goals of a full or satisfying life. Correspondingly, they posit a causal link between health and well-being. Whether well-being is a matter of the greatest balance of pleasure over pain (hedonism), fulfilment of desires (desire-fulfilment theory), or developing and exercising one's natural capacities (perfectionism), good health takes a prominent place among the large number of things that make a positive contribution to well-being (e.g., achievements, pleasure, friendship), and ill health among the things that detract from well-being (e.g., poverty, stress, discomfort). On virtually all accounts of well-being, pathological conditions are considered bad for us, as they typically involve pain, discomfort, distress, loss of capabilities, weakness, or fatigue. They are conditions that we *suffer* from, even if, in some cases, they can be outweighed by an accompanying good (e.g., asthma may excuse us from military service, but we would prefer to avoid military service without asthma, see Schroeder 2016). Experiencing pain or nausea detracts not merely from people's health, but also their well-being, while the alleviation of pain or nausea makes these people not only healthier, but also better off.

While accounts positing a causal link between positive health and well-being are consistent with the positive notion of health as considered in this chapter, some take a stronger position maintaining that health is simply a kind of well-being (see Hausman 2017). A prominent example is the WHO's definition of health, according to which health is a "a state of complete physical, mental and social well-being and not merely the absence of disease or infirmity."[14] Leaving aside the criticism that this renders health unattainable and difficult to operationalize (Huber et al. 2011), it is important to highlight that positive health is not a kind of well-being. There are at least two reasons for thinking so.

First, if health were a kind of well-being, then it would be hard to make sense of how it is possible for people to pursue well-being at the expense of health. For example, if a friend tells us that his high workload and excessive consumption of alcohol and cigarettes contribute to his well-being, we would probably caution him that his lifestyle is detrimental to his health. In doing so, we assume that health is not well-being.

Second, if health were a kind of well-being, then promoting health would necessarily involve promoting well-being. However, on a closer look, the relationship between health and well-being turns out to be less intimate. There are many examples in which a decrease in health (be it negative health or positive health) does not affect well-being. For example,

[14] Slightly less idealistically, Richard Kraut (2009, 90) seems to have something similar in mind when he maintains that "to flourish is simply to be healthy – to be an organism that is unimpeded in its healthy functioning."

a significant percentage of men over the age of sixty harbor prostate cancer, but die of other causes before the disease ever manifests itself. On both positive and negative conceptions, these individuals have decreased health due to a pathological condition, yet the condition need not decrease well-being. Correspondingly, increasing health by removing the pathological condition would not necessarily increase well-being.

In other cases, the decrease in health is perhaps more clearly manifest in a loss of capability due to a pathological condition, but it still has no effect on well-being because the pathological condition hampers a capability the exercise of which one disvalues.[15] Consider the following case:

> *Globozoospermia.* A 50-year-old man contacts his physician to get a vasectomy because he does not want to have more children. Before the intervention, by chance, he finds out that he has grown infertile due to *globozoospermia*, which is otherwise harmless and symptomless. Learning that he does not have a healthy reproductive system he is relieved that he does not need the vasectomy and the physician does not make recommendations that could attempt to restore his reproductive health.

This is a case in which health problems are outweighed by some accompanying good: if the 50-year-old had to choose between globozoospermia and a vasectomy to reach his goal, he would opt for the pathological condition, which, unlike the vasectomy, carries no further risks. The impact on well-being is thus contingent on what the agent values: whether one can contribute to his well-being by contributing to his health will depend on what he values. Such cases help illustrate that health is neither a necessary component of well-being, nor a kind of well-being.[16]

[15] In general, pathological conditions can affect capabilities, but whether they also affect well-being might depend on what the agent values. Somebody wholeheartedly dedicated to a vegan lifestyle need not experience a decrease in well-being by developing alpha-gal allergy (allergy to mammalian meat due to tick bite).

[16] The proximity of positive health to well-being can be rendered more intelligible if one accepts that "health" is not a concept with necessary and sufficient conditions for category membership, but a *prototype concept* with a probabilistic structure, in which membership in the extension of the concept is graded. Key concepts in the sciences (e.g., psychology, biology) have in recent decades been productively reconceptualized as prototype concepts, and the claim that "health" exhibits a prototype structure is plausible in light of the lack of a definition of the concept and the extensive area that it covers. This would mean that whether something qualifies as healthy is a matter of degree and hinges on grades of similarity to prototypical cases, but also that prototypical cases gradually shade into nonprototypes, which again gradually shade into nonmembers. Prototype concepts allow and often involve *overlap* such that an instance can simultaneously be a peripheral member of one concept and a more prototypical member of another, neighboring concept. This may perhaps explain why many have conceived of positive health as some type of well-being. Nonprototypes have few or no attributes in common, and they can exhibit attributes in common with members of neighboring concepts (e.g., well-being). (See Ramsey 1992).

6.6 Sanocentricity, Health, and Autonomy

This chapter started out with a very common idea according to which medicine is pathocentric. This was replaced with the suggestion that medicine is sanocentric: its aim is to promote health, but not necessarily by treating and preventing disease. This change allowed us to incorporate the vast majority of cases considered so far (i.e., cure, treatment, management, and prevention). We have also seen that there are two concepts of health and that opting for the positive concept allows us both to uphold the thesis that the aim of medicine is to promote health and to accommodate cases like age-related sarcopenia and pregnancy nausea. But while we have established that medicine is not pathocentric, but sanocentric, thus aiming to promote health, we have not yet provided a defense for the *Autonomy Thesis*, which not only states that the aim of medicine is to promote health, but also links the promotion of health to increasing autonomy.

But our opponent might argue that the introduction of a positive concept leads to a potentially devastating problem: since health is comprehended as a capacity, like resilience or robustness, it can always be enhanced, which means that on our account, medicine has no upper limit to permissible health promotion. This, however, would render our account overly permissive and inviting the charge that it cannot exclude cases of overmedicalization. In other words, we would reach an overly liberal comprehension of the aim of medicine on which virtually any intervention that promotes health is permitted. This would force anyone embracing the view proposed here to accept that much of the criticism of overmedicalization is misguided.

This objection is serious and reveals that our thesis hitherto that medicine is sanocentric needs to be adjusted. As a first step toward such adjustment, the next sections present four observations on the relationship between health and autonomy in the context of clinical medicine. Based on these observations, we will implement a final modification to the idea that the aim of medicine is to promote health and introduce the *Autonomy Thesis*. It will be argued that the objection can be defused, because the promotion of health, according to the *Autonomy Thesis*, is guided and limited by considerations about autonomy, which, according to a very rough, but generally accepted account, refers to the capacity to direct and determine one's actions in light of one's own principles.[17]

[17] A person is autonomous just in case she demonstrates self-governance, guiding her life from her own perspective instead of being manipulated or forced into a specific course of action by external

Promoting health is the proximal aim of medicine, pursued to the extent that it serves or is at least consistent with competency conditions (e.g., rational thought, self-control) and authenticity conditions (ability to identify with one's desires, values) (Berofsky 1995; Christman 2009). To understand this relationship, and to help comprehend how health is entwined with autonomy, the following sections make four observations about the relationship between health and autonomy in the context of clinical medicine. It is argued that, together, they suggest that the thesis that medicine is sanocentric needs to be adjusted.

(a) Autonomy Guides Considerations about Health Promotion

Considerations about what is best in terms of health are sometimes unable to solve conflicts between different treatment options and between treatment and prevention (e.g., in cases in which a treatment for one disease, like type 2 diabetes, can raise the likelihood of another, like heart failure). Making decisions in such cases only by reflecting about which option would result in the maximal increase of health is unlikely to provide sufficient guidance without considering what the respective health gains would mean for the autonomy of the individual. Here is a case to illustrate this point.

> *Elective amputation.* A patient with Body Integrity Identity Disorder (BIID) complains that despite repeated efforts to reconcile himself to living with his body, there is a profound mismatch between his phenomenal experience of his body (i.e., body schema) and the actual structure of his body. In particular, he stresses that his left leg has always felt intensely "alien" to him, which is why he requests that the leg be amputated. He has been evaluated by a psychiatrist who confirms that he is not suffering from delusions (e.g., voices telling him to seek amputation, believing that the leg was artificially added to his body) and exhibits full decision-making capacity.[18]

Whatever the appropriate course of action is in such a complicated case, making decisions only by reflecting on which option would result in the

factors (Christman 2009). There is not much agreement, however, as to how self-government is to be comprehended, and accounts vary depending on the theoretical and practical context in which the self-government occurs.

[18] There might be a number of reasons why somebody might harbor a desire for the amputation of a limb. Some people with Body Dysmorphic Disorder (BDD) request amputation in order to get rid of a limb that they incorrectly believe to be extremely ugly or affected by disease. Typically, individuals with BDD are resistant to evidence against their belief, which brings it close to monothematic delusions that consist in misperceptions of their bodies. In others, the desire for amputation is linked to sexual attraction to amputees or to wanting to become an amputee (apotemnophilia).

maximal increase of health is unlikely to provide sufficient guidance without considering what the respective health gains would mean for the autonomy of the individual. Assuming that the amputation would cure the BIID, is living with an amputated leg healthier than living with BIID? Instead, it is much more plausible to suppose that decision-making hinges on a reflection on the motive for seeking amputation and how the "alien" leg is impeding this man's autonomy. Such reflection could reveal that surgery could, all things considered, indeed augment his autonomy and relieve psychological suffering by enabling him to finally be who he feels he is supposed to be.

The point here is not to argue in favor of a particular course of action, but we should note that there is fundamental disagreement in the literature with respect to such cases. Some argue that with nonpsychotic and well-informed patients, the principle of respect for the patient's autonomy renders elective amputations permissible (Bayne and Levy 2005), while others think that it involves a violation of the integrity of the (healthy) human body that renders it inconsistent with the aim of medicine, at least to a degree that allows physicians to refuse surgery without violating the internal morality of medicine.

The main point of this brief discussion is merely to show that a reflection on what is healthier is unlikely to provide sufficient guidance without comprehending what it means for the autonomy of the individual. The course of action that will best promote the aim of medicine will in such cases depend on reflection not on health, but on autonomy. Simply put, how best to increase health becomes a question of how to increase autonomy, suggesting that the thesis that medicine is sanocentric needs to be adjusted. Of course, one might object that considerations about autonomy in such cases are just like other considerations about aspects other than health (e.g., cost-benefit ratio, the fair distribution of resources) that regularly play a role in decision-making. However, these types of considerations, unlike those about health and autonomy, are not internal to medicine.

(b) The Promotion of Health Is Not Permissible at the Cost of Autonomy

There is relatively broad agreement in contemporary medical ethics that competent patients who have the required capacities for self-government, are free from external constraints, and are sufficiently well informed should always be allowed to refuse treatment aiming to improve their health, even if leaving the condition untreated will lead to certain death. The promotion of health is constrained by considerations about autonomy: if a patient who fulfills these criteria decides against accepting the treatment, then respecting her autonomy dictates that she should not be treated against her will. Some might object that disobeying her decision can in fact amount to respecting her autonomy because treatment is the only way in which she will be able to continue living as an autonomous agent (see, e.g., Varelius 2005). Also, one might think that psychiatry offers a number of examples of medicine limiting autonomy for the sake of health. However, such interventions are only justified and consistent with the internal morality of medicine to the extent that the autonomy of patients is severely restricted, for instance by mental disorder. This is of course an oversimplified depiction of a complex debate, but fortunately, it is required neither to add more detail nor to take sides. What matters for our context is that on both views autonomy outweighs health, but does so without this somehow violating the aim of medicine or its internal norms. This also seems to indicate that the thesis that medicine is sanocentric needs to be amended.

(c) Autonomy Is a Legitimate Aim When the Promotion of Health Is No Longer Possible

In some cases, health promotion is no longer possible and biomedical understanding is deployed to explain and prognosticate the progression of a disease.

> *Ruptured aneurysm.* An older person whose life has been plagued by anxiety presents with an acute headache. Tests reveal that he has developed a type of ruptured aneurysm that cannot be treated. The physician in charge is able to completely relieve the pain, carefully explains the nature of the condition, and predicts that without any significant pain or discomfort, the person will fall into a coma in two to four days and die after another three to four. The physician also educates the partner about the condition and gives advice that might help cope with the situation, including about socioeconomic support.

In such cases, the prognosis does not appear to have import for the health of the patient. Moreover, given the overall psychological vulnerabilities of the patient, most accounts of well-being would deny that the prognosis actually increases the well-being of this person for the time he has left. Instead, medical understanding is used to predict the course of the disease and to promote the autonomy of the patient by offering him a chance to be in control over whatever is left of his life. When supporting autonomy via prediction in such cases, the physician is not pursuing some nonmedical aim. Correspondingly, choosing not to provide this information is not something that the physician could opt to do without violating the internal norms in medicine.

(d) Autonomy Is in Exceptional Cases Permissibly Pursued at the Cost of Health

Finally, a different sort of case involves the promotion of autonomy at the expense of health. In this regard, we may use a much discussed case in the literature involving sterilization in women.

> *Surgical sterilization.* A woman in her early forties contacts her obstetrician-gynecologist and requests surgical sterilization (tubal ligation) after having considered but declined other options for contraception due to possible side effects. She informs the physician that she does not want more children, and that her partner is in agreement with the decision. The physician has no reason to think that there is a risk for regret and helps her obtain the procedure.

In the literature, assuming a negative concept of health, there is broad agreement that this medical intervention does not target disease, but reduces health as it renders the reproductive system inoperative, and, less notably, exposes the patient to (low) risks generally associated with surgeries. On a positive concept of health, it is easier to see that sterilization can improve health, especially when considering that unwanted pregnancies can lead to an overall decrease in health, most typically in the form of short- and long-term mental health problems, including maternal depression (see McCrory and McNally 2013; Yanikkerem, Ay, and Piro 2013). Although not explicitly noted, this may be one of the reasons why the American College of Obstetricians and Gynecologists sees sterilization as a legitimate medical intervention that is consistent with the internal norms of medicine (American College of Obstetricians and Gynecologists 2017). If this is true, then such cases of surgical sterilization might be similar to the types of cases in which reflections on health are insufficient and considerations about autonomy enter the picture.

This line of reasoning also applies to contraception. Some accounts that operate with a narrower notion of health maintain that because fertility is not a disease, contraception and sterilization are "borderline" or "peripheral" medical practices: they are neither within the appropriate domain of medicine nor unambiguously supported by its aims (Brody and Miller 1998; Miller, Brody, and Chung 2000). These accounts accept that deciding whether and when to have children is significant with respect to the kind of life one wants to live, and that unwanted pregnancy can interfere with this. In other words, they accept that reproductive control can be of significance to autonomy. But while they accept that giving individuals control over their reproductive capacities may be a legitimate aim, they stress that it is not the aim of medicine. In contrast, on the positive concept, contraception is effortlessly accommodated and consistent with the currently accepted practice that decision-competent women may legitimately expect that their request for contraception will be met unless there are suitably powerful countervailing reasons.

A much less discussed but poignant case in which increasing autonomy is achieved at the price of decreasing health is the medical procedure in which a healthy organ is removed from the body of a healthy donor.

> *Removing an organ for donation.* A 27-year-old male signed up to donate a kidney to a stranger (nondirected or altruistic donation) through a non-profit organization. He reports that as soon as he learned about the pros of kidney donation and the low impact on the donor, he immediately knew what to do. He has always been deeply committed to the idea that if it is possible to alleviate suffering at a low cost for oneself, one should take action. Shortly after, specialists at a local transplant center removed the kidney and the donor returned to normal activities after four weeks.

In this case, it seems quite clear that, whatever benefit is gained by this medical procedure, it is achieved at the price of decreasing the health of the donor while being entirely consistent with the internal morality of medicine. Apart from the pain and (minimal) risks associated with the procedure, the donor faces a higher risk of developing certain conditions (e.g., high blood pressure and proteinuria). But note that here the anticipated effect on autonomy is very high: besides the autonomy of the recipient, the medical procedure also supports the autonomy of the donor. After all, it is consistent with his deeply held moral conviction and helps foster certain values that he takes to define who he is.

6.7 The Autonomy Thesis and the Overinclusiveness Objection

To recapitulate, introducing the positive concept of health into the thesis that medicine is sanocentric allows for accommodating plausible cases, but it also presents some problems. To start exploring how they can be addressed, we stressed that the promotion of health is guided and limited by considerations about autonomy. In order to comprehend how health is entwined with autonomy, we made four observations. Considered together, these observations help implement a final modification to the thesis that the aim of medicine is to promote health and also introduce the

Autonomy Thesis in its final shape. Observation (a) helps shed light on the importance of autonomy in cases in which the aim of promoting health does not offer clear guidance for decision-making. In such cases, which health goal ought to be pursued will depend on how the expected increase in health linked to a particular course of medical action would serve the autonomy of the individual. The other three observations carry a stronger weight in our argument, as they illustrate that (b) autonomy sets the limits of health promotion, and that (c) autonomy can be a legitimate medical aim, at least in cases in which health promotion is not an option. Thus, (b) and (c) suggest mutual restrictions: medicine can only pursue health as long as it does not undermine autonomy, and autonomy can only be pursued by medical means if health is no longer an option.

However, adding (d) to the mix poses an additional challenge: if cases like that discussed in (d), that is, the removal of an organ for donation, are legitimate medical interventions and consistent with the internal norms of medicine, then this generates a problem for the thesis that the aim of medicine is to promote health. If autonomy can permissibly be pursued at the cost of health, even if only in exceptional cases, then it cannot be true that the aim of medicine is to promote health.

To solve this problem, we implement a final modification to the thesis that the aim of medicine is to promote health. With this, we arrive at the *Autonomy Thesis*, according to which promoting health is the *proximal aim* of medicine, pursued to the extent that it serves or is at least consistent with the *final aim* of promoting autonomy. It is only in exceptional cases (e.g., the surgical removal of organs for donation) that biomedical understanding can be deployed to increase autonomy in a way that does not proceed via the proximal aim of promoting health. Having added this final modification, our thesis is now consistent with observations (a) to (c), and the problems raised by observations in (d) disappear. At the same time, the overinclusiveness objection loses its bite, because autonomy sets the limits of health promotion.

6.8 The Autonomy Thesis and the Second Overinclusiveness Objection

Our opponent may point out that a similar problem may resurface with the notion of autonomy. Indeed, some authors have argued that medicine's aim cannot be autonomy, because medicine would then become "merely an instrument to maximize individual choice and desire" (Callahan et al. 1996, 16). Our opponent could argue that since the

Autonomy Thesis allows that autonomy can in exceptional cases permissibly be pursued at the cost of health, this opens the door for highly controversial medical interventions to qualify as legitimate parts of medicine. Two such cases often discussed in the literature are prescribing anabolic-androgenic steroids for athletes to increase performance and performing certain forms of cosmetic surgical breast augmentation in women. The opponent may stress that there is widespread agreement in the literature that such interventions would not serve the aim of medicine, and argue that since the interventions would promote autonomy, proponents of the *Autonomy Thesis* need to clarify why these cases should not be classified as exceptional cases. If not, another version of the overinclusiveness objection threatens.

To deal with this objection, we may start by noting that what characterized the exceptional cases discussed in Section 6.6 was the combination of (1) a significant increase in terms of autonomy and (2) a minor reduction in health. But there are reasons for thinking that in the controversial cases under discussion, either (a) or (b) are not sufficiently met.

In the case of anabolic-androgenic steroids for athletes, it may be argued that (b) is not met to qualify as an exceptional case. The health risks of nontherapeutic steroid consumption in athletes are substantial and include increased risk of cardiomyopathy, atherosclerotic vascular disease, hypomanic or manic syndromes, decreased sperm motility, erectile dysfunction, menstrual dysfunction, and substance use disorder (for a review, see Kersey et al. 2012). Consequently, the health risks are considerably higher than in the exceptional cases discussed in Section 6.6.

In the case of cosmetic surgical breast augmentation, we may assume that the health risks are minor (although surgery requires lifelong further operations), but there are at least two reasons for thinking that (a) may not be met to a degree that renders the intervention admissible as an exceptional case. First, while the majority are satisfied with the outcome of augmentation surgery (Coriddi et al. 2013), there is little empirical evidence of long-term improvement in psychosocial functioning, self-esteem, or body image. With a lack of compelling evidence, it is not obvious that the relevant complaints are best approached with surgery instead of some other intervention (Sandman and Hansson 2020). But, in that case, it is not clear that breast augmentation in such cases would lead to a significant increase in terms of autonomy.

Second, simply assuming that the requested medical intervention would promote autonomy would rely on an overly crude notion of autonomy, which basically boils down to desire satisfaction. But such a notion would be unable to take into account that desires and intentions might be deeply held and yet still heteronomous, for instance, if they are unreflected impulses resulting from compulsion or from internalized social oppression. In this regard, feminist scholars have argued that the desire to seek surgical transformation to approach an idealized body type can be formed by oppressive gender norms. But, in that case, pursuing their realization may in some cases not only fail to promote autonomy, but may even be detrimental to it (for a discussion, see Davis 1991; Chambers 2008). To be clear, this is not to deny that some elective cosmetic surgeries can improve self-image and strengthen self-confidence in a way that increases autonomy. The point is merely that it cannot be readily assumed that meeting the request of persons seeking cosmetic breast augmentation will promote their autonomy to a degree that characterizes exceptional cases.[19]

Overall, the objection revealed the need for clarifying why such controversial cases do not belong in the class of exceptional cases that the *Autonomy Thesis* would allow. We found that while exceptional cases meet two criteria, the controversial cases under discussion do not. For this reason, they fail to produce serious challenges to the *Autonomy Thesis*. This helps show that the *Autonomy Thesis* does not result in an overly liberal comprehension of the aim of medicine. The conclusion is consistent with the majority position in the literature that these controversial interventions would not serve the aim of medicine. While the *Autonomy Thesis* is shielded from such objections, it can neither be reduced to the thesis (a) that the aim of medicine is to promote health, as long as this is in line with the autonomous will of the patient, nor (b) that the aim of medicine is to enable people to achieve their autonomously formulated goals.

In liberal democracies there are other practices and interventions aiming at increasing autonomy. Some of them, like public health, do so by enhancing health, for instance, in the case of providing free nutritious lunches in schools. But this does not create a problem for the *Autonomy Thesis*, as public health targets health from the perspective of populations, focuses on prevention, and can sometimes intervene at the cost of individual autonomy. In contrast, (clinical) medicine targets health from the perspective of individuals, applies a certain type of understanding, and is subject to particular internal moral norms that make it a social practice with a unique profile.

[19] This general idea also applies to other interventions such as "ethnic plastic surgery," which typically involves altering typical racial features to approach Caucasian beauty ideals.

6.9 Conclusion

Debates about medicine often proceed under the assumption that medicine has a well-established aim and what really demand attention are questions about determining the most suitable approach toward its realization. However, some of the criticism explored in Chapter 1 and the deliberations in this chapter indicate that the aim of medicine is far from clear, and there are a host of reasons why attaining more clarity on this issue would be productive. Among other issues, it could assist debates on the effectiveness of medical interventions and the prioritization of health care resources.

In order to help improve this situation, the main task in this chapter was to propose and defend an account of the *constitutive aim* of medicine. Acknowledging the dangers of unreasonably simplifying a diverse enterprise, it was maintained that on some level of abstraction, medicine is a coherent enough enterprise to have a constitutive aim that patients and medical professionals can appeal to when they fear that medicine is used to serve "alien" purposes, be it due to economic, political, or other reasons.

Drawing on our defense of the *Understanding Thesis*, it was argued that medicine has the character of an inquiry that aims to achieve specific forms of understanding. Approaching the question about the aim in tandem with considerations on internal morality of medicine, the chapter has elaborated what understanding pathological conditions in the service of human agency might amount to. Starting from the opening proposal that medicine is pathocentric, the chapter proposed and defended the *Autonomy Thesis*, according to which medical understanding aims to promote health, with autonomy being the final aim of health promotion. Shifting to a positive concept of health helped articulate the thesis and accommodate controversial cases. At the same time, it was argued that the positive concept does not render the account overly unrestrictive and does not allow highly controversial procedures as legitimate parts of medicine.

CHAPTER 7

The Aim of Medicine II
Current Alternatives

7.1 Introduction

This chapter continues our reflections on the aim of medicine, but is dedicated to exploring and critically engaging contemporary accounts from the literature. It deals with two types of approaches. While the accounts by Edmund Pellegrino and Alex Broadbent each identify an overarching aim, the chapter will also consider three "list approaches" that catalogue several aims, including the Hastings Center Report and lists put forward by Howard Brody and Franklin G. Miller, Bengt Brülde, and Christopher Boorse.[1]

Two methodological remarks are due. First, the sheer number of different accounts explored here means that their reconstruction will not be able to do justice to many of their details. Our somewhat myopic focus will be on examining to what extent these accounts are able to overcome or bypass the challenges encountered when defending the *Autonomy Thesis*. Second, subjecting these contemporary views to critical scrutiny is not merely an essentially adversarial procedure that marks the negative phase of philosophical work, but largely serves as a means to assist contextualizing the proposal presented in Chapter 6. By inspecting the most relevant aspects of these accounts in light of the challenges from Chapter 6, we are also strengthening our proposal by considering paths that the account presented in this book has not taken.

7.2 Edmund Pellegrino

Pellegrino (2001, 563) defines clinical medicine as "the use of medical knowledge for healing and helping sick persons here and now, in the

[1] The approaches have also been described respectively as "teleological" and "consensual" (see Schramme 2017a).

individual physician-patient encounter." He argues that medicine has a fixed nature, defined by serving the aim of "healing," which resists cultural, political, and social changes. "Healing" lays bare the etymological connection between health and some idea of "wholeness," and the idea is that an act of "healing" is an act aimed at assisting someone to regain "wholeness." Because health – the ultimate end for healing – is often not achievable, Pellegrino stresses the difference between curing and healing, and emphasizes that "healing" covers acts that aim to help restore psychological and physiological function and some sense of harmony.

> To care, comfort, be present, help with coping, and to alleviate pain and suffering are healing acts as well as cure. In this sense, healing can occur when the patient is dying even when cure is impossible. Palliative care is a healing act adjusted to the good possible even in the face of the realities of an incurable illness. Cure may be futile but care is never futile. (Pellegrino 2001, 568)

Correspondingly, Pellegrino (2001, 569; Pellegrino and Thomasma 1981; 1993) operates with a positive and broad notion of health. As the ideal end of healing, health is on his account "the good of the whole person," which is quadripartite (medical, personal, human, and spiritual good) and hierarchically organized. This means that in the clinical encounter, the good which must be served is not merely some narrowly construed "medical good" conceived of in terms of restoring normal functioning, but "the good of the patient as a spiritual being, i.e., as one who, in his own way, acknowledges some end to life beyond material well-being" (Pellegrino 2001, 570). While Pellegrino recognizes that physicians are not experts on all of the relevant dimensions, he stresses that the pursuit of the "medical good" has to harmonize with the other aspects of the patient's good, while upholding the moral priority of the highest good over the lower ones.

> Whatever the origin and content of one's spiritual beliefs, the three lower levels of good I have described must accommodate to the spiritual good. For example, blood transfusion might be medically 'indicated' for the Jehovah's Witness, abortion of a genetically impaired fetus for a Catholic, or discontinuance of life support for an Orthodox Jew. But in these cases, the mere medical good could never be a healing act since it would violate the patient's highest good. (Pellegrino 2001, 570–1)

Following this brief reconstruction, we proceed by taking a closer look at Pellegrino's thoughts on the nature of medicine and its internal norms. His claim that medicine has some fixed nature and aim that resists cultural, political, and social changes might strike us as problematic. Without denying a significant degree of historical continuity since Hippocratic times, it is also true that medicine's internal norms have been constantly reinterpreted and new norms have been generated in response to changes in the moral and social landscape of our societies. Consider that since World War II, moral norms in clinical medicine and medical research have undergone substantial changes, reflected, for instance, in new guidelines for truth-telling and confidentiality.

Another striking example is the change in the obligation of physicians to obtain informed consent, which has in recent decades become part of the internal morality of the profession. It is something physicians are bound by qua physicians. Of course, this is not to say that violations do not occur. The point is merely that if medicine had a fixed essence, as Pellegrino claims, then it would be hard to explain such changes. But to those who accept that as a part of society medicine is not immune to changes in society's general moral landscape, such changes are entirely intelligible and anticipated. The point is not that internal norms are completely determined by external morality, but that internal norms of social practices like medicine are not insulated from historical changes in the general moral landscape of the societies that they are parts of.

While Pellegrino stresses that being ill is a universal human experience and that the internal morality of medicine is entirely founded on simple characteristics of the relation between a person seeking help and a healer (e.g., benefiting the patient, promoting health), that relationship itself is not immune to changes in the reconfiguration of medical practice, patient needs, conceptions of benefit and health, etc. Whether a physician does her job well cannot be fully answered by mere reflection on simple and fixed characteristics of the relation between a healer and a person seeking help, but will depend in part on the patient's outlook and life projects that make sense within a larger framework of social norms that constitute the moral landscape of a particular society.

We may find Pellegrino's account of the internal norms of medicine too restrictive, but we may also suspect that his notions of health and healing are too permissive. At least at first, "healing" seems to exclude a large number of common interventions like contraception or sterilization, and it is hard to see how "healing" might be used to adequately describe the lifelong management of a chronic condition. But given that Pellegrino uses a broad notion of health and explicitly stretches healing to include "comforting" and "being present," this worry can be accommodated. However, on such a relaxed notion of healing and health, comprehended as a kind of well-being that even includes a spiritual dimension, the question is whether Pellegrino's account is able to place any limitations on what medicine can permissibly promote. While stressing the dimension of care is commendable, the consequence is that most issues that affect well-being and that could potentially be addressed by medicine become legitimate objects of medical attention. Pellegrino's view cannot avoid profoundly expanding the proper scope of medicine, and it also generates consequences that are hard to accept. For example, if health is well-being in a sense that includes a dimension of spiritual good, then somebody who has no disease but is spiritually out of balance is not fully healthy.

The general idea that what counts as proper healing has to be adjusted to the good of the patient is not foreign to the account suggested in Chapter 6. Indeed, the *Autonomy Thesis* stressed that autonomy, which is to an extent relative to the patient's commitments and projects, places a limitation on health promotion. However, if one allows that what counts as healing (i.e., acts with the aim of bringing health) is dependent on the spiritual values of the patient, then we end up with a notion of health that is to a much larger extent relative to the patient's ideas about life. The idea that violating the patient's highest good cannot count as healing would collapse health into well-being, leading to problematic consequences that we highlighted in Chapter 6.

To take one of Pellegrino's examples, imagine that a Jehovah's Witness has an accident and receives a lifesaving transfusion while in a coma because the physicians have no knowledge of his spiritual beliefs. Upon recovery, he might think that his "highest good" was violated, but it would be very odd if he thought that his health was somehow damaged. Alternatively, the same patient could have consented to the transfusion, temporarily bracketing his spiritual belief in order to save his life. In such a case, someone subscribing to Pellegrino's view would have to accept the unappealing consequence that the lifesaving act was not a proper healing act and constituted a violation of medicine's internal morality.

7.3 Alex Broadbent

While Pellegrino uses "healing" in a broad sense, as something that is not necessarily achieved by cure, in his book *Philosophy of Medicine*, Broadbent uses it in a much more restricted sense, interchangeably with "cure" (Broadbent 2019, 36). Not unlike Pellegrino, "healing" or "cure" is at the core of Broadbent's reflections on the aim of medicine, but unlike Pellegrino, "cure" is understood in contradistinction to treatment and care as an intervention that removes a disease.

Broadbent argues that the ultimate aim of medicine, and not just that of the Western medical tradition, is curing disease, while many other aims that are usually associated with medicine can only be regarded as aims in virtue of being connected to cure. Maintaining that the aim of medicine is cure means that (a) a significant number of medical activities have cure as the ostensible goal and that (b) society at large comprehends cure as a goal for medicine (Broadbent 2019, 41–2). Well aware of some difficulties that such an account would face with accommodating something as common as prevention, Broadbent adds an important aspect by maintaining that cure and prevention are parts of a larger goal (i.e., to bring about "that disease is absent and health present").

> The reason I see the goals of cure and prevention as part of a larger goal is that they are both in fact have [*sic*] the same effect if they are achieved – that is, the removal of disease. The only difference is when the intervention takes place . . . But the goals of curing and preventing are subsidiary goals to the same end; if you have either goal, then you have the larger goal of favoring health over disease. We do not have a word for that, but if we did, I would use that, and say that were the goal of medicine. Since we don't, I say that the goal of medicine is the conjunction of the goal of cure and the goal of prevention. (Broadbent 2019, 53)

Broadbent (2019, 35) distinguishes between the goal of medicine and "the core business of medicine," which is what medicine "actually does." One reason for introducing this distinction is to confront the apparent puzzle that we attribute progress and a very high social status to medicine while its track record of cures is far from impressive. As Broadbent (2019, 60) puts it, "if the core business of medicine were to cure the sick, then medical traditions, disciplines, practices, interventions, or practitioners that were unable to reliably cure the sick would not persist." Instead, the competence – the exercise of which constitutes medicine's core business – is providing explanations and predictions regarding health and disease. The conclusion is that medicine is fundamentally "an inquiry into the nature and causes of health and disease, for the purpose of cure and prevention" (Broadbent 2019, 64).[2] Curing requires understanding, but not vice versa, which allows us to infer that understanding is more fundamental than cure. Conceived of in this manner, medicine can make progress without an increase in curing ability (Broadbent 2019, 92).

While Broadbent's account skilfully deals with a number of problems that have haunted the literature, and while this book builds on many of his

[2] It should be noted that Broadbent uses "disease" and "illness" interchangeably when discussing the core business of medicine.

theses, there are some remaining problems that the account proposed here is better suited to handle. A minor issue is that the proposal risks obscuring that cure and prevention have different criteria of success: while a cure is only successful if it removes the disease, preventive measures are considered successful if they significantly reduce the risk of developing disease. Thus, cure and prevention can each be deemed successful without achieving the same effect.

A more important issue is that while the distinction between the goal and core business is of crucial importance to Broadbent's account, it is also potentially problematic. Drawing on an example from Chapter 4, we can make two points here. NATO's constitutive goal is the defense of its member states, but after many years of peace, NATO's core competence is now logistics and nation-building, and its core business is mostly in logistics, research, nation-building, peace-keeping, etc. The first point is that when deploying these competences, for example, hosting a conference on self-driving vehicles, there will be many cases in which NATO does not make progress toward its constitutive aim, but essentially does something else. The second point is that while NATO could not maintain its identity if it failed to do anything to defend one of its member states under attack, it could opt out of doing its core business (e.g., nation-building, hosting conferences) without violating its identity and the commitment to its constitutive aim.

With these two points in mind, we see that the distinction between the aim and core business of medicine can lead to some implausible results. Take a common case where a family physician explains to a patient that she suffers from a viral infection of the upper respiratory tract (common cold) and predicts that while the symptoms will probably worsen over the next two days, she should definitively be able to return to work within a week. Broadbent is right that the explanation and prediction (prognosis) that the physician offers is a part of the everyday business and results from the exercise of a core competence. But it is also clear that on Broadbent's account the activity of the physician does not pursue the aim of medicine, which he defines as situated at the conjunction of cure and prevention. Instead, in Broadbent's framework, explanation and prediction appear to be uses of medical understanding for nonmedical purposes, perhaps something like reassuring the patient that nothing major is going on, making a contribution to improving the patient's sense of control over her life, or something else. This conclusion is entirely consistent with the way in which Broadbent envisages medical activities like pain relief, which he thinks are uses of medical skills and tools, but do not themselves pursue the aim of medicine (Broadbent 2019, 46–51). But this conclusion nevertheless runs into problems familiar from Chapter 6. If we accept that explanation, prediction, and pain relief are uses of medical understanding for nonmedical goals, then we reach the counterintuitive conclusion that physicians could always opt out of these activities without violating professional norms.

Opening a gap between the aim and core business of medicine is sensible, and, beyond dealing with the puzzle that Broadbent is concerned with, it helps account for a number of controversial cases considered in Chapter 6. However, it is a high price to pay. Many everyday medical activities would no longer count as pursuing the aim of medicine, including treating itching dry skin with an ointment, contraception, rehabilitation, alleviating pregnancy nausea, and performing a vasectomy. Instead, they would count as using medical understanding for nonmedical purposes with the consequence that physicians could refuse assistance by citing professional norms. In light of these difficulties, it is worth stressing that without having to open such a gap between aim and core business, the puzzle dissipates if one accepts that the aim and the core business is to offer care with the aim of promoting health, perhaps loosely along the lines of the *Autonomy Thesis*. Opting for such a solution closes the gap such that the core business of medicine not only counts as consistent with its aim, but actually as making steps toward its realization.

Given the *Understanding Thesis* and the significant role that the notion of understanding plays in the account presented in this book, Broadbent's point that the core competency of medicine is understanding demonstrated by prediction is worth underlining. However, unlike the position defended in this book, it is not clear whether understanding in medicine for Broadbent amounts to *scientific* understanding. The answer appears to be in the negative, as Broadbent maintains that the purpose of medicine and the purpose of science are different: medicine is "an inquiry for a certain purpose, and its purpose distinguishes it from science (whose purpose I do not want to say more about than that it is different from medicine's)" (Broadbent 2019, 64).

7.4 The Hastings Center Report

Instead of trying to identify one overarching goal like Pellegrino and Broadbent, "list approaches" proceed by cataloguing a number of goals that medicine pursues. In Sections 7.4–7.7 we will explore four list approaches, starting with the Hastings Center's consensus report (Callahan et al. 1996). The authors offer a list of "four goals of medicine":

(1) the prevention of disease and injury and the promotion and maintenance of health

(2) the relief of pain and suffering caused by maladies

(3) the care and cure of those with a malady, and the care of those who cannot be cured

(4) the avoidance of premature death and the pursuit of a peaceful death.

These goals leave open a number of questions, and we will not dwell on a number of issues raised by critics (see Boorse 2016). There are, however, some aspects that are helpful to our inquiry. First, without adding further constraints, (1) renders medicine overly inclusive. Further, the promotion and maintenance of health not only includes the prevention of disease and injury, but goes well beyond it. If the promotion of health is not further specified, then it can be taken to include legal health protection, free warm meals in schools, and political measures to increase the number of ICU beds in a geographical location. But even if we limit the promotion of health to the prevention of disease and injury, this increases the proper objects of medical concern to include a vast array of things like seat belts in cars, emergency exits in lecture halls, and carbon monoxide detectors.

Unlike much of the literature, the authors use the notion "malady" instead of speaking of "sickness," "disorder," or "disease" as most of the literature does. The authors define "malady" in the following fashion (Callahan et al. 1996, 9):

> The term 'malady' is meant to cover a variety of conditions, in addition to disease, that threaten health. They include impairment, injury, and defect. With this range of conditions in mind it is possible to define 'malady' as that circumstance in which a person is suffering, or at an increased risk of suffering an evil (untimely death, pain, disability, loss of freedom or opportunity, or loss of pleasure) in the absence of a distinct external cause.

The addition of "external cause" excludes cases of suffering arising from war or violence. But without additional constraints, "malady" is exceedingly inclusive: the loss of freedom or opportunity can be caused by a large number of things like old age, baldness, lack of musical skills, or pregnancy, which would all qualify as maladies. Concerns about inclusiveness are exacerbated by the fact that the notion of health the authors deploy goes beyond the absence of malady. After suggesting that health is "invisible" in the sense that it is not something one usually notices, the authors define health in the following manner (Callahan et al. 1996, 20):

> by 'health' we mean the experience of well-being and integrity of mind and body. It is characterized by an acceptable absence of significant malady, and consequently by a person's ability to pursue his or her vital goals and to function in ordinary social and work contexts.

There are several issues to mention here. It is not clear how one can depict health as something that is phenomenologically "invisible," that is, not directly experienced unless it is damaged, while also defining it as a particular *experience* of well-being and integrity (Callahan et al. 1996, 9). The emphasis on health as based on something phenomenologically salient seems inconsistent with the idea that health is phenomenologically "invisible." Perhaps the authors could say that health is invisible in the sense that it only figures in the background of one's experience of the world, but even so, collapsing the difference between health and well-being also raises a number of issues briefly explored in our discussion of Pellegrino's account. It leaves unresolved cases in which experiences of well-being and integrity are had by people who have undetected but potentially lethal diseases, or when healthy people fail to experience well-being and integrity (e.g., if one is exhausted or has low-grade anxiety or gastrointestinal discomfort).

Given the broad definition of health that the authors operate with, the second part of (1) actually includes most of the other items. The promotion and maintenance of health definitively includes (3) (i.e., care and cure of malady), which already includes (2) (i.e., relieving pain caused by malady). As to (4), the first half (i.e., the avoidance of premature death) is clearly covered by (1), while the second half (i.e., the pursuit of a peaceful death) is included in (3) (i.e., caring for patients who cannot be cured). We should note that (4) highlights an important dimension of care, which involves "the empathetic and continuing psychological care of a person who must, one way or another, come to terms with the reality of illness. . . . medicine may have to help the chronically ill person forge a new identity" (Callahan et al. 1996, 13). But on the notion of health that the authors deploy, this qualifies as a part of health promotion and is thus covered by (1).

The authors acknowledge that some uses of medical knowledge are not compatible with the aim of medicine, and they divide what they see as potential misuses into four categories. First, activities that are unacceptable under any and all circumstances include participation in torture and capital punishment. Second, activities that are beyond the conventional goals of medicine but serve morally appropriate purposes include cosmetic surgery to improve appearance, contraception, sterilization, and abortion. Third, activities that could be acceptable under some circumstances include "the use of medical knowledge to enhance, or improve upon, natural human characteristics" (Callahan et al. 1996, 15). While it is not clear why cosmetic surgery to improve appearance belongs in this category, the authors offer examples of genetic and pharmacological enhancement such as administering growth hormone treatment for healthy but short children or using anabolic steroids for athletic enhancement. Fourth, under the label "uses of medicine unacceptable under all but the rarest circumstances" (Callahan et al. 1996, 16), the authors discuss whether confidentiality might be broken to enable identifying the bearers of viruses and performing contact tracing to protect the health of others. They warn about the dangers of using medical knowledge for manipulative and coercive purposes in the name of improved health (e.g., coerced abortions, screening, diagnosis, and forced change of unhealthy habits). This part of the account is difficult to evaluate, as it remains unexplained how breaking confidentiality in the case of a virus epidemic belongs to the same category as massive coercion and how the latter types of cases could be acceptable even in the rarest circumstances, if they pose a threat to the "institution of medicine and to human liberty and dignity."

In all, (1) covers all the other aims, which means that the list is reducible to a single aim. Moreover, the notion of health that the authors deploy goes beyond the absence of malady. It is tied to the experience of well-being, which is one of the reasons why the account is overly inclusive.

7.5 Bengt Brülde

Brülde's account of the aim of medicine is constructed around a discussion of the findings of the Hastings Center Report. Consistent with most approaches and with our use of the term, Brülde uses "medicine" in a broad sense that includes traditional hospital services, palliative care, and rehabilitation, but excludes medical research. Unlike Pellegrino and the authors of the Hastings Center Report, Brülde (2001, 3) construes health as a matter of functional ability, in which being healthy is roughly equated with functioning well. Having adopted such a view, he suggests that "we can regard the goal to promote health as a special case of a more general goal, namely the goal to promote the patient's functioning as a whole" (Brülde 2001, 3).

At first, one may suspect that promoting functioning as a whole is an aim that is too broad to be able to deal with controversial cases. How would the norm "functioning as a whole" apply to cases of elective cosmetic surgery and functional enhancement? But from the variety of possible construals of what functional ability can be, Brülde chooses a rather narrow one. Noting that not every improvement in functioning is also an improvement in health, he stresses that "there are abilities (e.g., like the ability to read or to swim) that have nothing to do with health" (Brülde 2001, 3). But if such abilities are not related to health, then promoting functioning as a whole becomes both too broad and too narrow. A consequence of maintaining that the ability to read is not linked to health, whether or not one thinks of health in negative or positive terms, is the implausible one that dyslexia and alexia are mistakenly classified as disorders.

More importantly for our context, Brülde raises an interesting point according to which whether or not one operates with a positive or negative account of health, the aim to promote health can be "eliminated," that is, reduced to something more fundamental.

> To the extent that 'health' is understood as freedom from disease, the health goal coincides with the goal to combat disease, and to the extent that 'health' is understood in terms of wellbeing, the health goal can be regarded as a special case of the more general goal to promote quality of life. In short, no matter what we take health to be, the goal to promote health can be eliminated as a separate goal. (Brülde 2001, 4)

It is helpful to think about the promotion of health as a special case of some more general goal (e.g., well-being, autonomy), and this is indeed in line with what the *Autonomy Thesis* suggests. However, that does not mean that it can be eliminated without consequences. Two issues are worth noting here. First, as the discussion in Chapter 6 confirmed, the promotion of such a more general goal by medical means without going through health would overstretch the boundaries of medicine and would make it difficult to identify cases of overmedicalization. Second, it is unclear how the goal to promote health can be "eliminated," as Brülde does in this paper, if health is equated with functional ability. Health is then not eliminated, only reintroduced as functional ability.

Brülde presents a list of goals of medicine that are taken to be irreducible to one other and that are further qualified in the course of the paper.[3] He maintains that the diversity of goals would be concealed by introducing some kind of overarching goal. The first four goals are "instrumental" in

[3] I will refer to the qualified version, in part because there are some discrepancies between the first list and the qualified one, for instance, the seven goals are compressed into six.

the sense that medicine should only try to realize these goals when they are anticipated to exert a positive influence on the length and/or quality of life. The first four goals are as follows.

(1) To promote functional ability (including health).

(2) To achieve normal clinical status (including the avoidance of disease).

(3) To help the patient to cope well with her condition, and to improve the external conditions under which people live.

(4) To promote optimal growth and development.

With respect to (1), Brülde adds that medicine should only aim at good functioning to the extent that this positively affects the quality of life, or when it is necessary for the continued "development" of the patient. But as Brülde conceives of quality of life in terms of well-being (Brülde 2001, 4), it is difficult to see how such an account could deal with cases in which a goal like prediction (prognosis) is being pursued but at the cost of well-being. It is equally difficult to see how the account could place limitations in cases in which people seek medical assistance for enhancing their abilities beyond normal functioning that would positively affect their quality of life.

As to (2), Brülde adds that this aim should only be pursued when it is anticipated to lead to improvements with respect to the quality and/or length of life. The crucial issue of what "normal clinical status" means is not defined, but since it cannot be more than health and functioning, (2) is reducible to (1). Regarding (3), Brülde specifies that good external circumstances in this context refer to circumstances that tend to have positive effects on functioning. But here also, unless further constraints are introduced, school lunches, gun control, and environmental protection come into the purview of medicine.

Similar to the case with (2), (4) is reducible to (1). Brülde (2001, 7) holds that "the goal of medicine is not optimal development *per se*; it should only be promoted if this is expected to have a positive effect on functioning and quality of life." But then, growth and development are relative to functional abilities covered in (1), promoted because of their positive effects on the quality of life.

In addition, Brülde presents two final goals:

(5) To save and prolong life.

(6) To promote quality of life.

Qualifying (5), Brülde immediately notes that it is strictly speaking not a final goal, and medicine should only aim at saving and extending lives to the extent that these are worth living. What lives those would look like is a question he explicitly – and prudently – leaves open. But regardless of the answer to this question, the more important issue is that, just like (1)–(4), (5) is instrumental to (6). Moreover, we may note that (5) can be condensed, as saving a life always entails prolonging that life, but also that (5) is actually reducible to (1): it does not seem possible to prolong a life without successfully promoting basic functional abilities.

Finally, concerning (6), Brülde acknowledges that the goal to improve the quality of life in all respects and by any means would be unreasonably extensive. But he does not qualify the notion of "quality of life," and proceeds instead by laying out open questions about (i) whether medicine should be restricted to the relief of pain and suffering, (ii) what means medicine may permissibly take (directly or indirectly), and (iii) what kinds of interventions or treatments may be undertaken. In the end, the final goal of medicine boils down to promoting quality of life, thus (6), while all the other goals can be conceived of as instrumental to this final goal. Moreover, goals (2)–(5) can be reduced to (1).

7.6 Howard Brody and Franklin G. Miller

Howard Brody and Franklin G. Miller (1998) provide a list of eight aims. Echoing Pellegrino, they maintain that the aim of medicine derives from "healing" and argue that medicine's aims "are directed to a variety of ways in which physicians help patients who are confronting disease or injury" (Brody and Miller 1998, 386). Interestingly, Brody and Miller acknowledge that medicine has a single, essential aim, when they note that "there exists a fundamental unity of purpose among, say, working in an intensive care unit, a long-term stroke rehabilitation unit, and a home-care hospice. In each of these settings the physician is dedicated to benefitting patients in need of medical treatment and care" (Brody and Miller 1998, 387). But, if the authors believe that there is an overarching aim uniting the broad range of activities that make up clinical medicine, one might wonder why they choose not to present it together with their list.

Brody and Miller (1998) maintain that a list approach (a) illustrates medicine's diverse activities and skills and (b) helps avoid that this single goal is interpreted too narrowly. They thus choose to offer a list for pragmatic reasons: a single goal is bound to be interpreted too narrowly as "curing," and the list helps draw awareness to aspects of medicine that risk becoming overlooked. However, it is important to stress that these are not reasons for thinking that a list approach is superior to a single aim approach. Brody and Miller reasonably worry that identifying one aim might lead to misinterpretations, but this does not mean that a single aim cannot be determined without necessarily leading to misunderstandings.

The authors note that the list is not complete, and maintain that the goals of medicine *include* these items (1998, 386–7):

(1) Reassuring the "worried well" who have no disease or injury;

(2) Diagnosing the disease or injury;

(3) Helping the patient to understand the disease, its prognosis, and its effects on his or her life;

(4) Preventing disease or injury if possible;

(5) Curing the disease or repairing the injury if possible;

(6) Lessening the pain or disability caused by the disease or injury;

(7) Helping the patient to live with whatever pain or disability cannot be prevented;

(8) When all else fails, helping the patient to die with dignity and peace.

Compared to the Hastings Center's consensus report, the dimension of care is more explicitly defined under (6) to (8), and understanding enters the picture in (3), even if indirectly, as the goal to help the patient understand the condition. More importantly, Brody and Miller's list does not catalogue promoting health as an aim, only preventing disease and injury and caring for people with disease and injury. This is consistent with aims (2) to (8), but not with (1). It appears inconsistent to maintain that the list captures "a variety of ways in which physicians help patients who are confronting disease or injury" (Brody and Miller 1998, 386) while also including in the list reassuring the "worried well" – thus (1) – which does not involve dealing with disease or injury at all. If we take (1) seriously, then medicine is not limited to disease or injury, and the list of aims could be reduced to one single overarching aim. It is, however, difficult to evaluate this point, because the authors do not offer a complete list, and suffice instead to say that the goals of medicine *include* the items on their list.

Some of the goals seem overly inclusive, others overly restrictive. As to the first type of problem, some of the goals on the list indicate that a wide variety of activities can count as pursuing the goals of medicine. For example, it is easy to see that (4) "preventing disease or injury if possible" could include health information campaigns, security measures in work environments, providing free-of-charge school lunches, or even preventing war by political means. As to the second type of problem, (6) stands out because "lessening the pain or disability caused by the disease or injury" seems to exclude the care of pain not caused by disease or injury, like menstruation cramps.

There are other indications that Brody and Miller's account may be too restrictive. Like many others, the authors link their list of goals to the internal morality of medicine, but argue that while some medical activities clearly violate the internal morality (i.e., treating family members, prescribing steroids for athletes, execution, and having sex with patients) there are also a number of "borderline" cases that internal morality does not forbid (Brody and Miller 1998, 390). The authors do not take a position on nonreconstructive cosmetic surgery, but maintain that a principled justification for medical activities offering contraception and sterilization is hard to provide and perhaps only acceptable to due to external, "practical" considerations such as the fact that physicians have obtained a monopoly over the relevant procedures (Brody and Miller 1998, 391).

But the talk of "borderline" cases is not an appropriate solution for the problem. First, one wonders how something can be permissible if it constitutes a "borderline" case. Instead, the label "borderline cases" suggests that it cannot be decided whether or not they would violate the internal morality of medicine. Second, note that these procedures are not consistent with the way that the list of goals is formulated. No goal on the list seems able to accommodate contraception and sterilization, so without further additions or clarification, the conclusion is that these do not pursue genuinely medical goals. This not only strengthens our first observation, but leads to a problematic consequence: if these activities or interventions do not pursue a genuinely medical goal, then they are neither prohibited nor mandatory, which means that the physician is under no (internal) obligation to help. She could refuse treatment by invoking the nature of the profession. This conclusion is hard to swallow, but it is also hard to avoid if one accepts Brody and Miller's account. But, as argued in Chapter 6, if X is the goal of an activity that one has "signed up for," then one may opt out of doing something with the goal Y, even if Y is consistent with X.

7.7 Christopher Boorse

The most comprehensive list approach is found in the work of Christopher Boorse, who offers a list of seven aims of medicine. Fusing some of the items on previous lists, the goals on Boorse's list are as follows (Boorse 2016, 170):

Goals of Benefit to the Patient

I. Preventing pathological conditions
II. Reducing the severity of pathological conditions
III. Ameliorating the effects of pathological conditions
IV. Using biomedical knowledge or technology in the best interests of the patient

Knowledge Goals

V. Discovering the diagnosis, etiology, and prognosis of the patient's disease, including its response to various treatments
VI. Gaining scientific knowledge about the patient's disease type and disease in general, including their response to various treatments
VII. Gaining scientific knowledge of normal body function

The first thing to note is the two-part structure of the list (i.e., benefit to patients and scientific knowledge) and the addition of "knowledge goals" in the second section, which remedies the omission in other lists of scientific aims. This is commendable, but distinguishing "goals of benefit to the patient" from "knowledge goals" appears misleading if we accept that what makes medical research properly *medical* is that, in the end, some health benefit to patients is envisaged and expected. In such cases, knowledge (or better, understanding) is pursued in the interest of promoting health. Boorse could, of course, deny this point, but only at the price of collapsing the distinction between biological and medical research.

The second issue to note is that the goals in both sections are covered by a single goal. As to the section "Goals of Benefit to the Patient," Boorse notes that the Hastings Center Report's "care" and Miller and Brody's goals (3), (6), and (7) are absorbed into III, while (1) is a part of IV, changing "disease or injury" to "pathological condition." Boorse also acknowledges that IV collapses goals I–IV into one goal: preventing pathological conditions, decreasing their severity, and moderating their effects are included in the aim of using biomedical knowledge or technology in the best interests of the patient. This covers pain relief (whether or not it results from disease or injury), contraception, cosmetic surgery, and obstetrical anesthesia. IV also specifies the scope of I, II, and III, fixing the conditions for when they ought to be pursued. Moreover, IV stresses that there is no duty to eliminate diseases at any cost (e.g., in cases where the treatment has worse effects than the disease), or even to diagnose them. As an example, Boorse mentions prostate-specific antigen (PSA) screening, which detects early prostate cancers, but may lead to treatment that is in many cases unnecessary and harmful.

But IV also reaches into the second section and covers V and VI. After all, benefiting the patient who suffers from a disease includes determining the diagnosis, etiology, prognosis, and treatment, which requires biomedical knowledge of diseases and their responses to treatment options. So, apart from VII, which may be treated as a separate goal, all the goals on Boorse's list are covered in IV.[4]

Compared to the other accounts, Boorse's list widens the permissible aims of medicine, and he advocates a permissive view on which medicine is not limited to treating pathologies. Boorse anticipates objections that his account reaches "an ultrapermissive view of medical treatment" (Boorse 2016, 173) and defends his view by discussing a number of historical and contemporary examples, which purportedly show that medicine is not restricted to dealing with disease (e.g., Victorian era obstetric anesthesia, contraception, sterilization, relief of discomfort caused by menstrual cramps and teething, sleep cycle adjustment to counterbalance jet lag, cosmetic surgery, treating typical dysfunctions of old age, organ removal for donation).[5] In light of all these examples, Boorse thinks that it would be wrong to claim that medicine is limited to treating diseases and that internal morality would be violated in such cases. More precisely, he argues that three positions are possible:

> One is to retreat, and reject as unethical all our examples of physicians' justified treatment of normal conditions. The second is to endorse these examples as ethical acts by physicians, but not medicine since not directed at health. The third is to accept them as medicine, embracing an IMM (integral morality of medicine, SV) containing, on the patient side, only the single goal of using biomedical knowledge and technology for patients' benefit. (Boorse 2016, 173)

Boorse sensibly rejects the first option, as it would exclude too many activities in medicine not related to pathology, including a number of

[4] The case with goal VII is not straightforward, and I shall remain neutral on whether it can be covered by IV. But note that gaining scientific knowledge of how the body functions might be too broad, because it dissolves the difference between research in biology (aiming to understand body functioning) and medical research (aiming to understand body functioning with respect to improving our grasp of diseases and health benefits to patients).

[5] McAndrew (2019) argues that some of the items on the list might actually turn out to be consistent with the view of medicine as pathocentric. He notes that contractions, but not the labor pains themselves, are a natural part of childbirth. Also, he notes that adjustments to sleep cycles might be treated as risk management, because shift working or jet lag might lead to pathological conditions like heart disease.

those listed in the previous paragraph. He continues by arguing that the choice is really only between the second and third options. The important point according to Boorse is that, whichever we choose, the result has no effect on controversial cases and on limitations for physicians qua physicians. The second option would hold onto the view that medicine is pathocentric, but maintain that physicians are not limited to treating pathological conditions, while the third option would deny that medicine is pathocentric, which would entail that physicians are not limited to treating pathological conditions. So, whether one opts for the second or third option, the objections based on the internal morality of medicine lose their bite. No internal morality of medicine prohibits controversial activities, and rejecting them must have its source in external moral ideas. As long as these interventions are in patients' best interests, thus consistent with goal IV, they are acceptable.

Boorse's account has a number of advantages over other list approaches, but it is incomplete in an important sense. To see why, we may start by pointing out that while Boorse suggests a choice between the two last options, his account in fact needs to opt for the third option. This is because, on the second option, physicians could opt out without violating their roles as physicians. They could legitimately refuse interventions that do not serve medical aims and do so by appealing to the nature of the profession, even if this might violate external moral norms. This consequence of taking the second option may work for some of the most controversial cases (e.g., enhancement in the case of the young athlete), but not in cases like the relief of pain and discomfort, contraception, and sterilization. It is hard to imagine how physicians could refuse to help in such cases without violating professional norms.

While Boorse's account requires choosing the third option, doing so makes it vulnerable to the charge of being overly permissive. It is easy to see that voluntary assisted euthanasia and physician-assisted suicide are allowed on his account as far as they benefit the patient, but Boorse also allows a range of problematic cases (e.g., cosmetic surgery without any pathological condition, steroids for athletes to improve performance). Boorse's account still leaves intact some general internal norms that limit medical action (e.g., it excludes participation in torture or the death penalty), but without explaining what "benefit" means it is hard to see how the objection that the account is overly permissive could be deflected. Boorse (2016, 174) adds the "the duty of doctors not to harm their patients, and their duty to use their own judgment in deciding what is harm," but also here, much will turn on how the opposite of benefit (i.e., harm) will be comprehended. Cases of enhancement especially illustrate the need for specifying by what standards benefit and harm are judged. Would an "ethnic plastic surgery" be of benefit for a minority patient? Would anabolic steroids benefit the young athlete? Or would they be harmful?

It appears that in order to make progress here, one cannot avoid spelling out "benefit" in terms of promoting health, well-being, or, as suggested in this book, autonomy. Boorse (2016, 174) maintains that such cases might not constitute a problem, because "for a long time, feasible genuine enhancements will be few," and there is "doubt that doctors will soon be able to be trusted to improve on normality." If true, this might make the problem less acute, but it does not contribute to solving it. While we may agree with Boorse that the case against enhancement is not very promising, we may still think that his defense against being overly permissive is not convincing.

As a response to Boorse's account, David B. Hershenov (2020) argues that it is possible to stake out an alternative position that holds on to the idea that medicine is essentially pathocentric. He distinguishes between acts antagonistic to medicine's "essence" of combating pathology (e.g., inducing pathology, abortion, executions, and torture) and those that are not contrary to this essence but merely "not entailed by it." Hershenov argues that, while medicine excludes certain activities that are contrary to its essence, it does not exclude actions not entailed by its essence. Medicine does not violate its identity and internal norms if it uses medical means for goals that are not contrary to its essence. The pathocentric essence provides grounds for refusing without penalty activities antagonistic to medicine's essence, but physicians are not restricted to actions entailed by the "essence" of combating pathologies.

The argument has structural similarities to Brody and Miller's account, and the distinction between core and noncore activities has similarities to their distinction between core and "borderline" activities. It also suffers from similar problems and generates similar counterintuitive conclusions. First, one could argue that the account is overly restrictive. Hershenov assumes that enhancements are neither contrary to nor entailed by medicine's essence, but this neglects that they often actually induce injuries and pathological conditions – just think about removing a kidney from a living donor, or surgical sterilizations, both of which damage normal functioning. Enhancements are controversial and demanding to accommodate exactly because of this feature. But if this is correct, then on Hershenov's account, enhancements, sterilization, and a large number of other interventions end up falling under activities that the internal morality of medicine forbids.

Second, while advocating that physicians are not limited to fighting pathologies, the position advocated by Hershenov remains undecided on cases like contraception or the relief of discomfort, nausea, or low-level pain without disease or induced by treatment. These are cases in which the physician could deploy her biomedical understanding to benefit the patient, but as in the case of Brody and Miller's account, the consequence is that the physician, qua physician, is under no obligation to help. Helping to relieve discomfort or pain and offering contraception are neither prohibited nor mandatory as judged by the standard of the internal morality, which means that the physician could refuse treatment by invoking the nature of the profession. Helping out would then be something like a charity-based or supererogatory action, which is optional, beyond any professional duty, and the omission of which would not justify criticism.

7.8 Conclusion

Following the defense of the *Autonomy Thesis* in Chapter 6, this chapter was dedicated to exploring and critically engaging a number of rival accounts that have reached different conclusions. In part because consideration was given to a relatively high number of accounts (two approaches that identify a single goal and four list approaches), the focus was not on doing justice to all of their features, but on exploring whether they are able to deal with some of the challenging cases that were discussed in detail in Chapter 6 and that any account of the aim of medicine faces. The objective in subjecting these contemporary views to critical scrutiny with particular emphasis on the challenges from Chapter 6 was to strengthen the proposal presented in Chapter 6 by considering alternatives that we have chosen not to take.

Overall, the accounts explored here were found unable to deal adequately with some of the challenges discussed in Chapter 6. Single aim approaches were found lacking with respect to accommodating controversial cases. To mention some issues, Pellegrino's notions of health and healing were found too permissive, while Broadbent's distinction between the aim and core business of medicine generated implausible results. With respect to the list approaches, we reached a similar result. Moreover, many of the listed goals could be collapsed into broader goals or comprehended as subgoals or means to achieve them, in spite of claims that list approaches are superior to single aim approaches.

CHAPTER 8

Rethinking the Challenges
The Moderate Position

8.1 Returning to the Criticism

The book started out by exploring some of medicine's current challenges. We have highlighted that the criticism is *comprehensive* (i.e., it targets medicine both as medical science and medical practice) and that it charges that medical science is less trustworthy as a science than generally thought (skepticism), that medical resources are inappropriately used to address nonmedical problems (overmedicalization), and that medical care fails to meet expectations in certain regards (objectification). Moreover, instead of condemning medicine by deploying independently justified standards, the criticism is *internal* – it appeals to both epistemic and moral norms within medicine, and maintains that medicine has diverted from its course such that its aim fails to be realized in current institutional settings.

It was argued that whether or not the criticism is justified depends on fundamental questions about (a) the nature of medicine and (b) the aim of medicine. In the course of Chapters 2–7, we have made progress on both (a) and (b). According to the *Systematicity Thesis* and *Understanding Thesis*, medicine is science, thus a systematic inquiry with the (epistemic) aim of attaining understanding. According to the *Autonomy Thesis*, the primary aim of understanding in medicine is to promote health with the final aim of promoting autonomy. These findings allow us in this final chapter to return to the criticism and the issues it stresses, and to rethink them in light of our three theses. Taken together, the three theses allow us to take up what we call the *Moderate Position* with respect to the challenges, which deserves its name because it in most cases it situates itself between more radical views.

The chapter proceeds by addressing the three strands of criticism one by one and reconsidering the challenges they present in light of the *Moderate Position*. The focus is on getting a better grip on what the criticism highlights as wrong-making features, assessing to what extent the criticism is justified, and productively using the conclusions that we reach to point toward solutions. Concluding the undertaking of this book, the chapter uses previous findings to contribute to the debate about the future trajectory of medicine, and it points to ways in which some current challenges could be remedied.

8.2 Skepticism, Systematicity, and Well-Orderedness

Condensing large parts of the skeptical criticisms, Richard Horton, the then chief editor of *The Lancet*, maintained that "afflicted by studies with small sample sizes, tiny effects, invalid exploratory analyses, and flagrant conflicts of interest, together with an obsession for pursuing fashionable trends of dubious importance, science has taken a turn towards darkness" (Horton 2015, 1380). Skeptical criticism of this sort voiced by prominent figures and disseminated in leading journals sent shockwaves through the scientific community. It has motivated a radically skeptical attitude that has been described as "medical nihilism" (Stegenga 2018), which stands in stark contrast to those who regard the quality of evidence in medical research with the kind of optimistic attitude that characterized the first phase of the evidence-based medicine (EBM) movement.

It is, however, both possible and prudent to assume a more *Moderate Position*, which takes these critical voices seriously and adopts a critical attitude (e.g., expect to see evidence suggesting that an intervention is effective even when it is not), but does so without embracing nihilism. The *Moderate Position* can agree that most medical interventions are unsuccessful and that the chances of new "magic bullets" that effectively target very specific causes of disease are slim. It may also concur that medical interventions, on average, have relatively small effect sizes, that the institutional structure producing medical evidence is biased in favor of positive evidence and of underestimating harms, and that current evidential standards do not entirely eliminate problems with the malleability of the research methods. However, the *Moderate Position* emphasizes that acknowledging the scale of these problems need not lead to nihilism: while the task is massive, there is nothing suggesting that significant progress toward ameliorating the situation is not achievable.

In addition, the *Moderate Position* suggests that steps toward improvement can be based on the *Systematicity Thesis*. The wrong-making feature of the issues that Horton raised is that they violate the nature of scientific inquiry and thereby impede making progress. Here, the dimensions of systematicity not only enable us to attain more clarity with respect to the problem, but they also assist in shedding light on how the situation could be improved by increasing systematicity. As indicated in Chapter 3, systematicity can be understood in several ways, which is why we proceed by briefly outlining what increasing systematicity would mean in a narrow fashion (i.e., by adjusting evidential standards), and in an extended fashion (i.e., by adjusting the priorities of medical research). To be clear, the main aim here is not simply to use the language of systematicity to reiterate problems and solutions with respect to skepticism, but to show that the *Moderate Position* has significant advantages compared to other, more radical positions. To achieve these aims, it is helpful to deploy the language of systematicity, as seemingly separate issues become intelligible as problems with systematicity.

8.2.1 *Increasing Systematicity in a Narrow Sense*

It is important to note that deficiencies in terms of lacking adequate systematicity in a narrow fashion, for instance, due to biases and fraud, are not unique to medicine. In fact, it has become common to speak of a veritable crisis across the sciences. According to a survey of scientists published in *Nature* (Baker 2016), around 90 percent of respondents agreed that there is some degree of crisis, and the majority believed that chief contributors to irreproducible research include outright fraud, selective reporting, *p*-hacking, publication bias, and growing pressures to publish. While the "crisis narrative" about science may in some cases be exaggerated and counterproductive, sustaining antiscientific movements and fostering cynicism in younger scientists (Fanelli 2018), the problems with the quality and integrity of research and publication practices that medical skepticism stresses certainly exist across a large part of the scientific enterprise.

That said, there are features specific to medical research that contribute to exacerbating such problems and have in part fueled the nihilist position. First, the assessment of medical interventions is particularly complex: many diseases have complex physiological bases, and interventions on physiological systems lead to multifaceted causal interactions that are difficult to control. At the same time, financial interests are particularly strong and promote, among other issues, biases, spin (misrepresenting results to appear more newsworthy), and selective disclosure related to study results, methods, and design. Second, taking steps toward solving these problems in medical research may be even more difficult compared to other scientific fields because they require instituting changes on a wider array of dimensions (e.g., the academic rewards system, research culture, research funding, incentive structures, transparent interest disclosures for researchers as well as journal editors, policies in institutions that host research). In addition, it requires the participation of more stakeholders, including not only researchers, institutions, and funding structures, but also the pharmaceutical industry, journals, and patient organizations (Ioannidis 2016).

The existence of features specific to medical research that contribute to exacerbating general problems in science should not, however, conceal the fact that the general norms of systematicity by which the situation could be improved are actually not out of reach. It is not a controversial claim that increasing systematicity would help prevail over current limitations in identifying and removing sources of error, bias, and fraud. A closer approximation of the ideal of systematicity on the dimensions discussed in Chapter 3 (in particular the dimensions of "defense of knowledge claims" and "critical discourse") would help elevate epistemological standards and offset methodological shortcomings of trials and tools assessing the quality of evidence and corrupting influences. To mention one example, to curb publication bias and *p*-hacking, and to assure that the assessment process (e.g., the defense of knowledge claims) can proceed in a systematic fashion, it is increasingly accepted that prior to collecting data, trials ought to undergo preregistration, which includes the prespecification of the primary outcomes to be measured. It is relatively apparent that such measures that require specifying all relevant criteria prior to embarking on a research endeavor would assist in resolving some of the noted shortcomings.

Of course, this is indeed a very modest, perhaps even platitudinous position that not many will disagree with, and it does not imply that such measures will be able to eliminate completely all biases: research methods are never faultless and insecurities are liable to be exploited for "spin" in the continued presence of powerful incentives. However, one important point is that the *Moderate Position* proposed here is still more optimistic with respect to progress than nihilism.

Referring to what we described as measures to increase systematicity in a narrow sense as "detail-tweaking," Stegenga (2018, 187) stresses that they are "insufficient to resolve the problems that motivate medical nihilism." One reason is that if the implementation of "detail-tweaking" successfully decreases the bias toward overestimating effectiveness, then it also decreases estimates of effectiveness, which might be expected to strengthen the case for medical nihilism (Stegenga 2018, 188). This is a fair point, but it raises the question whether the attitude motivated by accepting decreased estimates of effectiveness still deserves the label "nihilism." For this, it is helpful to invoke the difference between *first-order evidence* and *higher-order evidence* (see, e.g., Christensen 2010). While the former is evidence that bears directly on some hypothesis, the latter is higher-order evidence, that is, evidence about evidence. Much of the force of the nihilist position comes from highlighting a tension with respect to evidence in medical research. Nihilism argues that while research offers first-order evidence that supports a certain hypothesis (about a medical intervention being effective), higher-order evidence about the malleability of the methods by which first-order evidence is gathered should make us be less confident in our judgments based on our first-order evidence. The tension is thus that we should have low confidence in the effectiveness of medical interventions even if our first-order evidence points to the contrary.

Now imagine that we have succeeded in increasing systematicity by implementing a substantial amount of "detail-tweaking," perhaps with the help of stricter regulatory measures. In such a case, we would not only have diminished bias, error, and fraud, but improved the epistemic situation in general by resolving the tension between first- and second-order evidence. It would no longer be the case that we are not justified in believing what our first-order evidence suggests. We would be forced to accept that many interventions are less effective than previously assumed, but we would be in a qualitatively different epistemic situation that would vindicate a more *Moderate Position* than nihilism.

An additional reason why the account offered here points toward a more *Moderate Position* is linked to the *Understanding Thesis* and the *Autonomy Thesis*. While Stegenga focuses on interventions using pharmaceuticals to target diseases, the two theses allow us to accept that progress in medicine also occurs when health care professionals are able to offer more precise prognoses (even if they cannot offer better treatments), better pain management (without the pain necessarily being linked to a disease), more patient-centered care, improved understanding of illness (as opposed to merely disease), better lifestyle interventions that promote positive health and autonomy, and so on. Unlike in Stegenga's framework (2008, 15), on the *Moderate Position* a medical intervention can qualify as effective without targeting the constitutive causal basis of a disease or the harms caused by the disease. For this reason, the *Moderate Position* does not exclude vaccination and interventions on a large number of other conditions (e.g., contraception, abortion, relieving teething pain or menstrual cramps), which means that its overall assessment of medicine will be more optimistic than that arrived at by the nihilist position. So, while the *Moderate Position* is more optimistic, it is both because it has a more positive opinion about the prospects of "detail-tweaking," and also because it allows a larger number of medical interventions to potentially qualify as effective.

8.2.2 *Increasing Systematicity in an Extended Sense*

If the epistemic aim of inquiry were merely to increase understanding, then increasing systematicity in a narrow fashion (i.e., adjusting the standards by which theories, hypotheses, findings, and observational practices are assessed) would be sufficient to ensure that research in medicine remained on track toward its aim. However, as argued in previous chapters, the aim of science is to increase understanding that is *significant*. This means that, taken together, the *Systematicity Thesis* and the *Understanding Thesis* allow us to maintain that scientific inquiry can be deficient both by lacking systematicity in a narrow sense, and also by lacking systematicity in an extended sense, for example, by insufficient systematic reflection at the agenda-setting stage on the significance of the anticipated gain in understanding.

Such lack of systematicity in an extended sense is not about failing to properly consider something like alternative measurement procedures or feasibility, but about failing to apply systematic reflection on what the envisaged increase in understanding would mean in a larger societal context. Once we accept the *Systematicity Thesis* and the *Understanding Thesis*, including the feature that the norms of systematicity extend to cover research goals and questions about significance, then consequences ensue for the choice of scientific endeavors. With respect to medicine, once we factor in the *Autonomy Thesis*, we see that increasing systematicity in the extended sense has important consequences and offers impulses for rethinking the priorities of medical research toward a better distribution of resources. Let us briefly explore three consequences.

First, extending systematicity requires that the medical research agenda be realigned to correct the skewed focus on drug interventions at the price of investigating strategies that improve health by nondrug or nonsurgical interventions (e.g., regular exercise, eating habits, social life, and stress management). If there are indications that such interventions have a substantial effect on risk management or disease outcome, then the requirements of systematicity would dictate devoting the necessary resources to fully investigate such possibilities. There are many other possibilities to mention. For example, some studies indicate that features of the built environment in hospital settings can improve patients' experiences and exert a positive impact on their health outcomes (for a review, see Jamshidi, Parker, and Hashemi 2020). Others suggest that empathic communication by primary care physicians can not only improve patient satisfaction and increase quality of life, but also reduce pain and mortality, and increase clinical safety (see Howick et al. 2020). While there are relatively few high-quality studies in these areas, increasing systematicity would require dedicating more research resources to assess the potential of possible interventions.

Second, extending systematicity would require taking steps toward correcting the tendency to let considerations about earning potential be the sole guide in devising research trajectories. For example, this would involve taking measures to counteract the focus on so-called me-too drugs, that is, drugs that belong to the same therapeutic class and are structurally very similar to an existing first-in-class compound, but differ only in some relatively marginal respects. These drugs are often lucrative without much added innovation and therapeutic advantage over their predecessors, although they increase the availability of alternatives in case of drug shortages and can add some improvements (e.g., once-a-day administration, fewer drug–drug interactions). However, with over 60 percent of the drugs listed on the WHO's essential list belonging in the category of "me-too drugs" (Aronson and Green 2020), research efforts channeled into such drugs are excessive and stifle exploration of new avenues that are scientifically promising, but do not appear very profitable. The current state of research on antibiotics serves as a good example here.

In an article in *Nature*, Maryn McKenna (2020), speaks of a "bitter paradox" with respect to antibiotics. Whereas antibiotics were responsible for a large part of the growth of many leading pharmaceutical companies, there is not much interest in research on new antibiotics because the earning potential is limited.[1] There are still relatively few resistant infections that require treatment with new antibiotics, and if they do, treatment usually does not exceed a few weeks, which means that the earning potential is limited. Moreover, as the development of resistant bacteria is

[1] McKenna (2020, 338) describes a rather dire scenario: "For almost two decades, the large corporations that once dominated antibiotic discovery have been fleeing the business, saying that the prices they can charge for these life-saving medicines are too low to support the cost of developing them. Most of the companies now working on antibiotics are small biotechnology firms, many of them running on credit, and many are failing."

taken more seriously, "antimicrobial stewardship" programs now promote the cautious use of antimicrobials, to reduce resistance and decrease infections caused by multi-drug-resistant organisms. This ensures the reliability of the antibiotics in the long term, but puts constraints on profits.

Third, extending systematicity in this manner and acknowledging considerations about significance as a proper part of science puts us on a path toward what Kitcher (2001; 2011; Flory and Kitcher 2004) has called "well-ordered science." According to Kitcher, science is well ordered if its inquiries are directed toward questions that are significant in the sense that they would be endorsed in a democratic deliberation among well-informed participants. The participants are in Kitcher's account an ideal collection of people who represent a wide range of different points of view and who are prepared to take seriously the interests and needs of the other participants. A crucial point is that the collection of participants cannot be restricted to people whose points of view only represent the affluent parts of the world. Instead, they must be chosen from behind the veil, thus without knowing whether they are members of wealthier societies. In this manner, one avoids the pitfall of letting the course of scientific inquiry reflect historical accidents by which some people came to live in wealthy societies that channel resources into science.

Currently, medical research is far from well ordered. Private and semi-private actors that fund most medical research tend to support endeavors that promise significant profits because they target wealthy populations. This means that they focus on diseases that mostly affect people in the affluent world instead of those that typically affect people in economically less fortunate areas of the world (e.g., typhus, malaria). More precisely, 90 percent of the global resources dedicated to medical research investigate diseases affecting 10 percent of the world's population.[2] Some of the criticism found in medical nihilism offers additional support to these concerns: while disproportionately large amounts of resources are being dedicated to the health problems of wealthy populations, the pharmaceutical products that emerge from this endeavor tend to be of low efficacy (Stegenga 2018, 190).

Well-ordered science requires that the needs of inhabitants of poor regions of the world be taken into account, and this means a change in research priorities, urging more research into, for example, infectious

[2] Based on a 1990 report from the WHO's Commission on Health Research for Development, the expression "10/90 gap" was introduced to highlight the discrepancy between the allocation of research funding and disease burden.

disease. Reiss and Kitcher (2009) argue that medicine could grow into a well-ordered science by adhering to a simple, "fair-share" principle, which requires that the global amount of suffering and harm caused by a particular disease be proportionate to the quantity of resources dedicated to it.

Kitcher's framework of "well-ordered" science is a helpful tool to identify and address problems related to the prioritization of resources dedicated to medical research, and the idea of extending systematicity in the manner proposed in this chapter points in a very similar direction. Similar to Kitcher's conclusion in light of the idea of "well-ordered" science, increasing systematicity as suggested here would require a significant modification in how research resources are distributed, correcting the disproportionate focus on diseases mainly affecting the wealthy parts of the world. There is, however, at least one major difference. While Kitcher's account assesses the direction and priorities of medical research in light of a moral standard that is external to science (i.e., the "fair-share" principle), our account only makes recourse to norms that are *internal* to it. A similar difference emerges with respect to Stegenga's suggestion (2018, 187) that medical research should shift its attention away from the "magic bullet" model and that medical research priorities should be changed to dedicate more resources to examining "gentle" interventions. This means that while the *Moderate Position* proposed here points toward some of the same types of reforms that Kitcher and Stegenga highlight, it is based on assessing a practice against its own standards and therefore does not face the task of having to demonstrate the legitimacy of applying an external standard to assess medicine.

Overall, based on our theses, the *Moderate Position* holds that increasing systematicity in the narrow and the extended senses offers ways to address the challenges raised by the skeptical criticism and provides input for reconsidering the priorities of medical research. According to this position, increasing systematicity would not merely lead to more reliable knowledge and understanding, but also help medicine raise and address problems in a way that enables it to fulfil its purposes and societal function.

8.3 Overmedicalization

To recapitulate, overmedicalization (i.e., the improper use of medical resources to address nonmedical problems) is medicalization gone wrong. While medicalization is a descriptive term, with respect to a condition X, it can occur in two ways. X can be medicalized with or without *pathologization*, thus with or without X attaining the label of a pathological condition. Medicalization thus describes a broader phenomenon, involving a shift toward comprehending various types of medical interventions as justified with respect to the condition (e.g., pregnancy, fertility, death),[3] whereas pathologization describes how certain conditions that enter the medical jurisdiction become labeled as pathological (e.g., alcoholism, epilepsy) (see Sholl 2017).

Corresponding to this distinction, critics argue that overmedicalization can occur with or without pathologization, but in both cases, it violates the proper aim of medicine. While the criticism is usually not supported with a detailed account of what that aim is, examining typical cases in the literature in light of the *Understanding Thesis* and the *Autonomy Thesis* allows us to offer more clarity on overmedicalization. Also here, our theses support the *Moderate Position*: as medicine is not limited to disease, medicalization is only improper overmedicalization if biomedical understanding is deployed in a manner that fails to promote health as outlined by the *Autonomy Thesis*. Let us consider both paths to overmedicalization in more detail.

8.3.1 Overmedicalization without Pathologization

In the most-discussed cases that belong to this category, critics object to pregnancy, fertility, and death coming under the purview of medicine. It is, however, often unclear what the wrong-making feature of overmedicalization is in such cases. There are several possibilities. One possibility is that overmedicalization is wrong simply because medicine extends its domain to include conditions previously beyond its boundaries. But there are two problems with this initial idea.

The first problem is that because medicalization also involves the extension of the domain of medicine, unless further modifications are added, this criticism collapses the conceptual space between medicalization and overmedicalization. This would be unacceptable, as it would turn every instance of medicalization into overmedicalization. The second problem is that, as it stands, the criticism seems to assume that the domains of social practices are somehow fixed. However, if social practices function, as Sally Haslanger (2018) argues, to coordinate our behavior

[3] Thusly understood, medicalization also includes vaccination campaigns as well as sanitation and public hygiene measures that target living and working conditions.

around resources, then it is natural to think that they can reasonably modify their domain and change in ways that are responsive to the changing demands that historical developments place on societies.

If these considerations are on the right path, then the mere extension of domain cannot constitute the wrong-making feature of overmedicalization. In light of our theses, the crucial question is not whether a state or condition historically belongs to the domain of medicine, but whether applying medical resources can offer an increased understanding of it. If it cannot, then the extension of domain would indeed be unwarranted. But clearly, this does not seem to be the case with pregnancy, fertility, or even death. If medicine provides adequate tools for understanding the relevant biological and psychological processes and if deploying them to these conditions can make a contribution to promoting health and autonomy, then, unless other reasons speak against it, the extension of domain is justified.

Many critics would not be swayed by these considerations. Taking a slightly different route, some critics maintain that the wrong-making feature is not so much that medicine extends its domain, but that this extension brings states and conditions that are not diseases into the purview of medicine. But this line of argument only works if one accepts the underlying assumption that medicine is limited to diseases. However, as argued in Chapter 6, adopting such a narrow conception of the permissible aim of medicine faces insurmountable challenges. So, if it is correct that medicine is not limited to disease, then the mere extension of domain that brings conditions that are not diseases into the purview of medicine cannot constitute the wrong-making feature of overmedicalization.

A better option for critics is to claim that the wrong-making feature with respect to the medicalization of pregnancy, fertility, and death is not the extension of domain in itself, but the fact that it leads to a loss of autonomy to medical authority. Take, for instance, the example of pregnancy, which, as critics rightly note, has increasingly become medicalized, that is, treated as a medical event that benefits from risk management by medical professionals. Pregnancy now typically involves regular contact with clinics and hospitals, and standard interventions offered include progress assessment based on ultrasound screening, blood tests, and various forms of genetic testing. Among other measures, pregnant women are informed about the effects of eating and drinking habits, stress, weight gain, and exposure to potentially toxic products. While the vast majority of women opt for medically managed pregnancies and carefully follow risk-minimization advice, critics highlight that the medicalization of pregnancy comes with expectations that pregnant women will regulate and monitor their bodies and their behavior consistent with standards established by medical professionals (for a discussion, see Kukla 2005). Overall, the crux of the criticism seems to be that the wrong-making feature of overmedicalization consists in (a) undermining self-determination by medicalized social control while (b) failing to offer health gains.

If this were an accurate description of medicalization, then anyone accepting the *Autonomy Thesis* would have to concur that the charge of overmedicalization is justified. On the account offered in this book, reducing autonomy while failing to promote health would clearly constitute a violation of the aim of medicine. It is, however, not at all clear that medicalization in these cases involves a loss of autonomy and fails to promote health. Let us briefly examine the purported wrong-making feature.

With respect to (b), it is hard to deny that medicine provides adequate tools for understanding biological and psychological processes in pregnancy, and that deploying such understanding allows making predictions about its likely course. This can help the pregnant woman increase control over her life by allowing her to make informed decisions about maternity leave, returning to work, and so on. Medical tests and control procedures can certainly lead to increased anxiety about what might go wrong, but this seems a relatively minor issue compared to the extent to which medicalization diminishes the risk of complications and unnecessary suffering (e.g., glycemic control during pregnancy reduces stillbirths and related complications like congenital anomalies and macrosomia; blood pressure monitoring reduces the risk of preeclampsia-related complications).

If women who choose a "medicalized pregnancy" attain more control over their lives, reduce their exposure to pain and risk of complications, and perhaps even increase their chances of assuming their regular professional or other activities sooner, then, on the *Moderate Position*, the medicalization of pregnancy does not qualify as overmedicalization and is entirely consistent with the aim of medicine as proposed by the *Autonomy Thesis*. The matter is of course different if it can be shown that there are medical interventions that have no benefits and involve added risks for patients.

As to (a), the charge of loss of autonomy to medical authority, we may start by exploring the question how this is different from the norms developed and advocated by those who favor nonmedicalized, "natural" pregnancy. Note that "natural" pregnancies are also overseen by experts. The difference is that the relevant experts are not medical doctors but midwives: they are health care professionals who, besides offering expert care during labor and delivery, also perform examinations and monitor early stages of pregnancies. Midwives do not operate outside of medicine; they often practice under the supervision of a physician or have a collaboration agreement with a physician. Given that midwives are medical professionals recognized as authorities on pregnancy and birth, whatever loss of authority to medical experts the critics claim occurs in medicalized pregnancies, it can also occur in nonmedicalized ones, even if they are not situated in an institutional setting.

It is true that the medicalization of pregnancy comes with expectations toward pregnant women and that the relevant norms are developed by medical professionals. But it is not clear how this is different from the norms developed by those who favor nonmedicalized, "natural" pregnancy. The idea of a "natural" birth has its own, often idealized expectations of how a pregnancy and birth ought to proceed (i.e., vaginal delivery free of pain medications and labor-inducing drugs), which can exert pressure on women to conform to them. Moreover, unlike in the case of medicalized pregnancies, a successful "natural" birth is often perceived as a sign of maternal qualities, which may induce a sense of having failed in women who experience complications during labor and need pain medicine or medical intervention (Lyerly 2012; Malacrida and Boulton 2014). Consequently, both medicalized and "natural" pregnancies can function to enforce a set of social norms, and both can destabilize the sense of control of pregnant women and subordinate them to expert authority, producing an estranged detachment from the process. Medicalization in itself can thus not be the wrong-making feature. A "natural" birth with a minimal amount of medical intervention can be empowering, but so can monitoring during pregnancy, ultrasounds and genetic tests, pain medication during delivery, or a scheduled cesarean.

In conclusion, the mere extension of domain that follows from redefining a condition or state as a problem in medical terms can only qualify as overmedicalization upon the addition of auxiliary claims about medicine, for instance, that it is limited to disease or that medicalization involves a loss of autonomy to medical authority. But upon closer scrutiny, these claims prove flawed. One might suspect that the idea that medical authority passively victimizes people is tied to a picture of authoritarian and paternalistic medicine that proceeds at the cost of the self-determination of individuals. While this may have been an accurate verdict, at least in parts of medicine in the 1950s and 1960s, when an early version of the criticism of overmedicalization first emerged, it does not convey an accurate picture of the situation today. In any case, in contrast to restrictive views that equate the mere extension of domain with overmedicalization, the *Moderate Position* proposed here is that if the medicalized condition picks out something that causes or increases the risk of suffering, harm, or death, and if medicine has the most satisfactory resources to understand it, then the charge of overmedicalization is not warranted and the condition may legitimately enter the purview of medicine.

8.3.2 Overmedicalization with Pathologization

The case is perhaps different when it comes to instances of allegedly improper medicalization that involve pathologization. The most-discussed cases of this sort include adult attention deficit hyperactivity disorder (ADHD), chronic fatigue syndrome, male pattern hair loss, hypoactive sexual desire disorder, mild cognitive impairment, obesity, Lyme disease, and fibromyalgia. The wrong-making feature is not merely that these conditions are medicalized, but that they are brought into the domain of medicine by being mistakenly classified as diseases. Before applying our findings to consider overmedicalization by pathologization, there are some important differences that need to be taken into account.

Typical cases adequately described as "disease-mongering" (i.e., the intentional construction of bogus disease categories or relaxation of diagnostic thresholds to increase the number of people who can be diagnosed) are fairly straightforward to assess. When motivated by monetary gain, by disapprobation of forms of behavior that do not conform to certain values of contemporary culture, or by the medical profession seeking to extend its domain, pathologization violates the aim of medicine as laid out by the *Autonomy Thesis*: it involves deception and fraud, does not promote health, and increases the risk of overtreatment. Moreover, reinterpreting a condition as a disease might lead to social disadvantages and stigma that are detrimental to autonomy.

In most cases, however, the situation is more complex, for example, when pathologization is propelled by individuals or patient organizations who seek to secure the best type of assistance for a condition. In numerous cases like chronic fatigue syndrome, individuals actually struggle to get diagnosed because receiving the diagnostic label helps qualify them as eligible for specialized medical assistance and a range of social services. In addition, a diagnosis might reduce moral blame and stigma and exculpate individuals from responsibility for having the condition. In such cases, pathologization renders suffering legitimate in the eyes of society and eases the burdens on the individual in a way that can promote their autonomy.

In cases that do not involve deliberate "disease-mongering" and fraud, once all the aspects are taken into consideration, the *Moderate Position* does not claim that pathologization violates the aim of medicine as proposed by the *Autonomy Thesis*. Instead, the view is that if the condition picked out qualifies as a problem and if medicine offers the most adequate understanding of it, it may not matter all that much whether it is classified as a disease. Of course, this is not an uncontroversial point. After all, when a condition is mistakenly categorized as a disease, a cognitive error of miscategorization occurs, which may be argued to constitute a wrong-making feature. In addition, it may decrease systematicity, at least on the dimension of description as described in Chapter 3. Such an error could thus hinder efforts to increase systematic understanding and ultimately to promote health.

In reply, the *Moderate Position* can point out that this may be a less serious issue for two reasons. First, with a lack of widely agreed upon criteria for what constitutes a disease or a pathological condition, debates on whether condition X is a disease are often unproductive and end up discussing the merits of different sets of inclusion criteria. In such a situation, if possible, it is prudent to suspend judgment. If this is not possible, then judgments about X should not be based mainly on ontological considerations, but rather on practical considerations about the benefits and disadvantages for patients of classifying X as pathological. Second, whether or not X is really a disease, or merely a condition that legitimately belongs to the purview of medicine and that medicine aims to provide a better understanding of, has relatively little influence on research and clinical practice. Of course, this is not to deny that this difference matters in other regards.

If we accept these two points together with the claim that medicine is not limited to pathological conditions (along the lines of the *Autonomy Thesis*), it looks like it is possible to largely sidestep difficult discussions about whether or not X is a *genuine* disease. Assuming that X causes or significantly increases the risk of suffering, harm, or death, whether or not X's entering the medical jurisdiction is consistent with the aim of medicine will depend not on getting the metaphysics right, but on pragmatic considerations having to do with consideration about whether medicine provides the most suitable means for understanding X.

8.3.3 Overmedicalization: The Moderate Position

Based on the *Autonomy Thesis*, the *Moderate Position* holds that the medicalization of a condition X does not amount to overmedicalization as long as (a) X causes or significantly increases the risk of suffering, harm, or death and (b) medicine offers an adequate way of understanding X that is superior to nonmedical ways of understanding it. Regardless of whether X qualifies as a disease, if conditions (a) and (b) are fulfilled, the charge of overmedicalization is unjustified, and medicalization will likely contribute to increase the ability of patients to understand and control their condition.

Importantly, whether or not medicine can offer a cure or effective treatment for X is not decisive: if employing medical understanding can help bring relief by giving a name to upsetting symptoms and by offering explanations and predictions, then this is engaging in the kind of (mental) health and autonomy-promoting activity that is entirely consistent with the *Autonomy Thesis*. In contrast, overmedicalization occurs when X enters the purview of medicine (a) without qualifying as a genuine problem (i.e., X does not cause or significantly increase the risk of suffering, harm, or death) and (b) without medicine being able to offer something in terms of understanding X. Independently of the question whether such overmedicalization also sidelines other ways of explaining, describing, or dealing with X, on our account of the aim of medicine, critics would in such a case be justified in denouncing medicine for overstepping its proper boundaries.

The *Moderate Position* advocated here might strike one as overly permissive. It is therefore important to stress that it does not deny that a number of serious problems related to medicalization remain. But these are issues that are worth treating separately, as they are more about how medicalization is being put to use and are not directly relevant to questions about whether medicalization violates the nature and aims of medicine. Let us briefly review five problems of this sort.

First, it is a problem if the medicalization of X completely crowds out other ways of explaining or describing X. For example, alcoholism fulfills conditions (a) and (b), and few would claim that it is an instance of overmedicalization. Nonetheless, even if medicine provides the most adequate means of understanding it, this does not warrant ignoring institutional, cultural, and social factors (e.g., class, images of masculinity, attitudes toward drug use) that influence the prevalence of alcoholism.

Second, while the medicalization of X can contribute to shifting the focus of problem-solving relevant to X to individual-level medical interventions, it should not make us ignore the relevance of interventions on political and social structures that often play a significant role in generating X or conditions like X. For example, medically treating conditions that are mainly caused by polluted or stressful work environments should not detract from political initiatives that aim to decrease the exposure to pollution or stress at work. When X enters the domain and jurisdiction of medicine, the fact that medicine might offer the most adequate means for *understanding* X does not mean that the question whether medicine offers the best means of *dealing* with X should be taken off the table.

Third, the fact that X comes into the purview of medicine by attaining the label of disease should not obstruct public deliberation about values in contemporary culture that might have played a role in singling out X as a problem. Such deliberation might lead to demedicalizing certain conditions (e.g., the removal of homosexuality from the DSM), and allow recognition of a larger natural variation in organism functioning, skills, and behavior.

Fourth, medicalization expands the category of what is perceived as demanding medical attention, and can thereby contribute to the increase of health care costs. This increase can lead to a wide range of problems, but as long as the medicalization of X leads to a better understanding of X in a way that promotes health and autonomy, its contribution to increasing costs does not represent a violation of the nature and aim of medicine.

Finally, it has often been pointed out that medicalization can lead to overtreatment. For example, it has been argued that screening programs deploying improved methods to detect early-stage prostate cancer have led to overtreatment. Many of those accurately diagnosed with prostate cancer would not develop the symptoms if left untreated, which means that they might receive unnecessary treatment associated with risks and significant side effects. But here also, the problem is not medicalization per se, but that the expansion of the category of what is perceived as demanding *medical attention* (i.e., having the pathophysiological basis of a disease) has led to the expansion of the category of what requires *therapeutic intervention*, in spite of this being associated with risks that do not clearly outweigh health benefits. The issue is that X entering the purview of medicine, as a disease or not, does not determine whether X requires therapeutic intervention.

8.4 Objectification

To recapitulate, the charge of objectification is that, in some cases, medical care discounts the personal experience of illness and its psychological and social dimensions in a way that is inconsistent with the nature and aim of medicine. Objectification is normally not the result of malicious intent, but rather the unintentional product of functional requirements in health care settings to which technological mediation and deindividualization can be identified as chief contributing factors.

First, increased dependence on sophisticated technology predisposes physicians to seeing the body of the patient in a way that the patient as a person risks "disappearing" in the encounter: illness experience is perceived as a veil that risks obscuring the physician's direct access to the underlying pathological condition. Second, the deindividualized appearance of patients in health care environments might make them appear less as agents that command empathy. This, combined with practices that can further contribute to collapsing the distance between the person and the disease (e.g., labeling patients in terms of their illnesses and medical procedures) and the anonymous uniforms of medical professionals, helps disguise the special responsibility that physicians have vis-à-vis patients as persons in vulnerable states.

Whether or not objectification violates the norms of medicine depends on what one believes to be the aim of medicine. If one thinks that the aim of medicine is merely to cure disease, then the charge would be unjustified, as objectification in the manner described here would not necessarily obstruct curing a disease. If one thinks that the aim is something like promoting well-being, then the charge would be justified and objectification would indeed be widespread. However, if we accept the *Autonomy Thesis* along with the *Understanding Thesis* and its implications in medicine, we are led to the *Moderate Position*. To see what this amounts to, let us briefly reiterate the role of personal understanding.

It was argued in Chapter 5 that understanding the condition of a patient requires a systematic inquiry that not only involves the application of biomedical understanding, but also a participatory personal understanding, which involves adopting a particular second-personal stance and deploying cognitive resources in addition to those required for biomedical understanding. We have distinguished between minimal and extended personal understanding and argued that a systematic inquiry in the clinical interview requires not merely understanding *disease*, but understanding specific instances of *illness* that reflect the individual's distinctive perspective on her conditions.

Once the crucial role of personal understanding is clarified vis-à-vis the application of biomedical understanding, and once it is clear that biomedical understanding is not sufficient for clinical understanding, it is not difficult to see that objectification can hinder the pursuit of the aim of medicine. More precisely, objectification can lead to harmful outcomes in at least two ways.

First, by hindering systematic inquiry, objectification can obstruct the promotion of health. We have seen in Chapter 5 that it can be important to be attuned to patients' facial expressions, postures, gestures, and abrupt changes in topic, not only because it leads to the disclosure of more comprehensive histories and allows for a more accurate diagnosis, but also because it improves the quality of the therapeutic relationship and leads to better compliance with proposed therapies. Objectification hinders such an attuned interaction: it obstructs personal understanding and stimulates forgetting that what is treated is not a disease, but a patient. Perceiving the patient less in terms of a person with a particular perspective and more like a biomechanical system with potentially impaired component parts hinders adopting the kind of engaged second-personal attitude that is required for personal understanding. Moreover, once minimal personal understanding is disturbed, attaining extended personal understanding becomes near impossible. It becomes impossible to take up the patient's perspective and thereby attain a comprehension of how the illness and the projected therapeutic outcomes align with her overall scheme of values, desires, and beliefs. Extended personal understanding, which includes perceiving the patient as a being similar to oneself, as someone whose life is centered on values worth caring about, does not get off the ground if minimal personal understanding is disturbed.

Second, objectification can have a detrimental effect on autonomy. We have argued that the promotion of health is only justifiably pursued if it is at least consistent with the final aim of promoting autonomy along the lines of the *Autonomy Thesis*. Serious and chronic illness can have significant negative effects on autonomy and agency, and in order to come to terms with and adapt to the condition, patients receiving such diagnoses will often be forced to rethink the values and desires that constitute the cornerstones of their identities. The process of adjusting is, however, rendered more difficult if the medical care they receive conveys to them the impression of being a "case." This may prevent or make it more difficult to be able to perceive oneself as a unique, valuable individual deserving careful attention, which makes the task of rebuilding a sense of personal identity even more demanding.

In conclusion, the *Autonomy Thesis* helps clarify the problem and on how internal norms in medicine can be violated, and how objectification hinders inquiry and understanding and can be detrimental for the promotion of health and autonomy. But while we see how objectification can hinder the pursuit of the aim of medicine, our examination of systematicity and biomedical understanding also helps us see the essential function of entities and processes associated with objectification (e.g., standardized roles, technologically sophisticated devices, decomposition into mechanistic subsystems). Many radical criticisms of objectification in medicine seem to leave little room for such features in clinical settings related to standardization, and they fail to see that even some forms of empathy reduction and moral disengagement for clinical problem solving can display functional aspects. In contrast, on our view of the aim of medicine and on the *Moderate Position* proposed here, some of these problematic features can be part of delivering effective care and fit the aims of medicine. So, the *Moderate Position* suffices by noting that medicine will fail to reap the full benefits of standardization and technological advances unless it implements measures that counteract the diminished personal understanding to which they contribute. While a detailed catalogue of what could be done is beyond the aims of the chapter, and would optimally involve empirical studies in clinical settings, our account helps identify what parts of these features could be changed without jeopardizing medicine reaching its aim.

Some practices that contribute to objectification and create unnecessary distance between health care personnel and patients could be corrected without detrimental effects. For example, personal understanding in both minimal and extended forms is disturbed by the deindividuating distance supported by practices such as wearing uniforms that indicate distinct group membership. Moreover, unnecessary distance could also be minimized by attaining a level of diversity in the population of health care professionals that reflects patient demographics. Some similar effect could be achieved by highlighting personal information about the patient (i.e., beyond name and the type of intervention) prior to consultations or interventions involving several health care professionals. Such measures could have effects similar to those that occur upon the attachment of facial photographs during radiograph interpretation, which has been shown to increase the rate of detection of simulated wrong-patient errors (Tridandapani, Bhatti, and Wick 2019) and to lead to more accurate diagnoses (Turner and Hadas-Halpern 2008). Although this study failed to replicate (see Ryan et al. 2015), the authors behind the replication attempt note promising differences that require further examination. Again, none of this is to deny that suppressing empathy can surely be useful when having to inflict pain or facing solving multifaceted clinical problems (Cheng et al. 2007). However, there is no evidence in the literature that such psychological processes could not occur without or need support from features of the clinical environment that are also associated with objectification.

For a final example, some of the standardized language in charts and discharge letters could be changed to better assist the overall aims of medical interventions. Clearly, standardized language is important to secure coordination across clinical contexts, but current medical parlance includes phrases like "the patient is medically optimized" (Dawson 2021) that are problematic in several ways. First, while being "medically optimized" means that no further medical intervention is planned or required, a patient who is immobile after having suffered a stroke may think that "medically optimized" is an inadequate (and perhaps also insensitive) description of her state. Second, the phrase also suggests that once "optimalization" has been achieved, medical care has reached its aim, implying that the rehabilitation measures that the patient with stroke will likely receive are supplementary to and not a part of achieving the aim of medicine. Third, as Dawson (2021) points out, it could promote a tendency toward clinical apathy and passivity on the part of the patient, hindering taking co-ownership of recovery. On its own, the impact of relatively minor matters like using the phrase "the patient is medically optimized" is of course very limited. But in combination with the other problematic features mentioned here, it is not difficult to see that it could run counter to the aim of medicine as proposed in this book.

8.5 Conclusion

Our investigation was guided by the idea that medicine's current challenges point to fundamental questions about medicine and that a systematic exploration of these will not only help address the challenges, but also offer productive input for rethinking the future course of medicine. Earlier chapters found that the criticism is comprehensive, internal, and depends on assumptions about fundamental issues in medicine. Having defended three distinct theses about these fundamental issues (i.e., the *Systematicity Thesis*, the *Understanding Thesis*, and the *Autonomy Thesis*), the main tasks of this chapter were to explore the three strands of criticism in light of these theses, to discuss the relevant wrong-making features, and to point toward possible solutions. The three theses supported what we called the *Moderate Position*, mainly because with respect to many of these issues, it can be situated between overly pessimistic and optimistic positions.

Rethinking skepticism in light of the *Systematicity Thesis* helped apprehend how certain structural features contribute to the violation of the epistemic norms of systematic inquiry that are internal to medicine qua being a science. It was argued that increasing systematicity in simple and extended senses offers ways to address the issues that skepticism stresses, yet without embracing nihilism. Bringing into play the *Understanding Thesis*, we have provided support for the claim that increasing systematicity also offers inputs for reconsidering resource allocation in medical research. Increasing systematicity would thus not merely lead to more reliable knowledge, but would also help prioritize research goals that are consistent with the aim of medicine.

Reconsidering the charge of overmedicalization in light of the *Autonomy Thesis* exposed that medicalization does not amount to overmedicalization as long as the relevant condition (a) causes or significantly increases the risk of suffering, harm, or death, and (b) medicine offers an adequate way of understanding it. If these conditions are fulfilled, then medicalization is consistent with the aim of medicine regardless of whether the condition qualifies as a disease and regardless of whether medicine can offer effective treatment for it. While medicine would not be overstepping its proper boundaries in such cases, a systematic understanding of a condition requires that medicalization does not crowd out other ways of explaining, describing, or dealing with that condition.

Finally, exploring the charge of objectification in light of the *Autonomy Thesis*, we found that it can hinder the pursuit of the aim of medicine in two ways. It can do so by hindering personal understanding, which thwarts systematic inquiry and can obstruct the promotion of health. Moreover, the absence of minimal personal understanding makes an extended form of personal understanding near impossible, which can have a detrimental effect on autonomy. Objectification can decrease one's ability to perceive oneself as a person worthy of care and support, which renders the task of coping with disease and maintaining a sense of personal identity even more arduous.

Conclusion

Disorientation and the "Greatest Benefit to Mankind"

C.1 "Doom and Gloom"?

In his groundbreaking work on the history of medicine, *The Greatest Benefit to Mankind* (1997), Roy Porter vividly describes the unparalleled ascent and expansion of medicine during the nineteenth and twentieth centuries. He portrays some of the forces (i.e., scientific advances, the socialization of medicine, and medicalization of society) that enabled its rise from a craft based on bedside observations to a respectable branch of science, and in the final chapter he notes that medicine is increasingly attracting criticism that expresses a fundamental *disorientation* about its aim and societal role. As he puts it, "medicine has become the prisoner of its own success. Having conquered many grave diseases and provided relief from suffering, its mandate has become muddled. What are its aims? Where does it stop?" (Porter 1997, 717). Porter does not offer an answer, but he closes the book by expressing serious concerns about the further expansion of medicine as leading to medicalization propelled by consumerism and inflated expectations.

Today, a quarter of a century after the publication of Porter's book, the tendencies Porter highlighted have become more pronounced. First, the forces behind medicine's expansion continue to exert an undiminished influence in the twenty-first century. While this is not the place to discuss the COVID-19 pandemic, the unprecedented measures undertaken to control it have in all probability increased the intensity and scope of these forces. The scientific advances that resulted from these efforts are said to have paved the way for future diagnostic and therapeutic interventions. The nature of the disease has brought into sharp focus the need for publicly funded research and centralized health care services (Galvani et al. 2022).[1] During the crisis, many noted that medicine has continued

[1] Speaking to a US context, Galvani et al. (2022) argue that a universal single-payer health care is fundamental to pandemic preparedness.

to expand its reach into private and public areas of life, leading to criticism from parts of the political spectrum that were hitherto not typically associated with debating medicalization. For example, although many of the relevant claims involved serious distortions, the resistance to mitigation measures in antiscience and personal liberty movements was sometimes framed as fighting the "medicalization of freedom" (Ezell 2022).

At the same time, the criticism of medicine has intensified since the publication of Porter's book. We have started our investigation in this book by drawing attention to three forms of criticism that stand out, each highlighting challenges to contemporary medicine. Together, the criticisms not only claim that medical science is less trustworthy than generally thought (skepticism), but also charge that medical means are increasingly deployed to nonmedical issues (overmedicalization) and that the care patients receive is in many ways inadequate (objectification). At the same time, figures like Raymond Tallis have objected that the "doom and gloom" of criticism paints a distorted picture of medicine. As he puts it in *Hippocratic Oaths*, medicine "has never been in better shape. Its scientific basis, the application of this science in clinical practice, the processes by which health care is delivered; the outcomes for patients, the accountability of professionals, and the way doctors and their patients interact with each other – all have improved enormously even during my thirty years as a practitioner. Yet the talk is all of doom and gloom: short memories have hidden the extraordinary advances of the last century" (Tallis 2004, 3).

In reply, two issues are worth highlighting. First, focusing on criticism is not inconsistent with acknowledging the developments in medicine or even with agreeing that when it comes to scientific quality, therapeutic efficacy, and the humane dimension of care, medicine, as Tallis (2004) puts it, "has never been in better shape." Second, Tallis seems to overlook something that Porter highlights, namely that our expectations for medicine evolve with the development of medicine. But if that is the case, then recalling the extraordinary advances of the past will not defuse the criticism.

Overall, it seems safe to conclude that the general tendencies noted by Porter have become even more pronounced in the years since his book was published. In some ways, the disorientation with respect to the aim of medicine that Porter diagnosed has worsened, and the sheer scope and depth of the criticism and the problems it highlights suggest that medicine's scope and role in society is fated to be altered in the twenty-first century. The fact that health care expenditure in developed countries has increased much more rapidly than gross national product will make an additional contribution to speeding up this process. At the same time, we have not made significant advances with respect to addressing the criticism and ending the disorientation. We have not attained considerable clarity on the question about the aim of medicine, with which Porter ended his investigation. At this critical threshold, confronting such questions could assist a constructive deliberation about medicine and influence its future course. This, in a nutshell, was one major motivation behind this book, combined with the hope that it might inspire further work at the intersection of philosophy and medicine.

Without pretending to have offered a comprehensive account of the challenges to medicine, the claim was that taking seriously and offering a detailed analysis of the current forms of criticism of medicine could offer key contributions to the ongoing deliberation. There are certain risks associated with devoting a relatively significant amount of attention to the criticism. Perhaps inspired by Tallis's remarks, some might resist this approach altogether because it risks contributing to an unconstructive "doom and gloom" discourse about medicine. Indeed, this is a point that cannot be dismissed out of hand, especially if we add that an unnuanced proliferation of depictions of medicine as being in a deep crisis brings with it a risk of becoming self-fulfilling: for example, it may lead to declining interest in funding medical research and diminishing trust in medical care. For this reason, the investigation aimed to avoid contributing to some crisis narrative, while taking seriously the criticism, driven by the persuasion that, upon suitable examination, it could offer valuable assistance in advancing the debate on medicine.

C.2 The Main Points

One important finding was that the criticism is not merely comprehensive (i.e., targeting medical science and medical practice) but also internal: it claims that medicine fails to meet *its own* internal standards and points to fundamental questions about (a) the aim of medicine, (b) the nature of medicine, and (c) the key concepts of health and disease. Clearly, a systematic philosophical examination of these fundamental questions carries the potential to assist in improving the current disorientation and deliberation about the future of medicine as a science and clinical practice. Chapters 3–7 were mainly dedicated to addressing these fundamental questions, while the overall framework in which they were approached was outlined in Chapter 2. Corresponding to the questions that the criticism raises, the approach distinguishes between *three levels of analysis* that are all required to address them. In contrast to an influential tradition in the philosophy of medicine, the endeavor is defined as a second-order philosophical inquiry that is *continuous* with normative elements that are more or less explicitly present in medical science and clinical medicine.

This formed the basis of the book's pursuit of three main goals, which were achieved by integrating separate discussions, such as new developments in epistemology and philosophy of science, as well as debates on the internal morality of medicine. The first two goals were to offer accounts of the *nature* of medicine and the *aim* of medicine to help improve a situation that Porter has described as characterized by disorientation. Based on these two accounts, the ambition was to return to the challenges to medicine conveyed by the criticism, rethink the critique, assess to what extent it is justified, and then outline possible solutions. With respect to reaching the third goal of the book, these accounts of the *nature* and *aim* of medicine helped assume the *Moderate Position*, which provides a better comprehension of the challenges, points toward possible solutions, and helps rethink the proper boundaries of medicine and the appropriate use of medical resources. The position earned the label *moderate* because it positions itself between more pessimistic and more optimistic views.

The overall theoretical approach having been presented, the focus was first on tackling questions about the nature of medicine. Chapter 3 addressed the general question about the nature of scientific inquiry with special attention to medicine. Against a prominent view holding that medicine is something other than science, the *Systematicity Thesis* assumes that systematicity is a necessary condition for science and that it generates reasoning and inquiry that produce reliable knowledge and understanding. On this basis, Chapter 3 illustrated how inquiry in medical science *and* clinical medicine meet the requirements for systematicity on nine different dimensions. Concluding that there are convincing reasons to accept the *Systematicity Thesis*, Chapter 3 showed that it is also helpful to comprehend what differentiates medicine from activities widely recognized as pseudoscience. Due to its lack of synchronic and diachronic systematicity that characterizes scientific inquiry in medicine, homeopathy is susceptible to a variety of biases.

Turning to the questions about the aim of medicine, Chapter 4 focused on the *epistemic* aim of inquiry in medical science. It defended the *Understanding Thesis*, according to which inquiry aims at *understanding*, leaving the question of what special kind of understanding there is at stake in medicine to subsequent chapters. Also here, we broke with an influential view in the philosophy of medicine, which maintains that due to its practical orientation, inquiry in medicine differs *in kind* from scientific inquiries. While showing that this view misconstrues the aims of scientific inquiry, in part because scientific inquiry is in more or less direct ways linked to the development of human agency, Chapter 4 recognized and highlighted key differences *in degree*, which help comprehend what counts as progress in medicine. One additional task was to underscore the implications of the *Understanding Thesis* for thinking about responsibilities in scientific inquiry. Linking these considerations to our findings in Chapter 3, these implications for responsibility were elucidated, showing that systematicity extends to deliberations about the choice of an inquiry.

The *Systematicity Thesis* and the *Understanding Thesis* together point to the following rough proposal. Assuming that practical interests are indexed to epistemic interests and that pathological conditions tend to undermine agency, the (epistemic) aim of medicine is to understand pathological conditions, which serves the final objective to be able to intervene on them in a way that contributes to the endeavor of supporting human agency. However, this rough proposal leaves unclear *what exactly* the character of understanding is in medicine and *how exactly* medicine contributes to supporting human agency. Chapter 5 mainly focused on the first issue, while Chapter 6 was dedicated to the second.

Chapter 5 drew on the epistemology of understanding to distinguish types of understanding in medicine and reflected on pivotal moments in the history of scurvy to explore what understanding a disease involves in the context of medicine. The hypothesis here is that, in inquiries in medical science, *objectual understanding* of a disease (i.e., biomedical understanding) requires grasping a mechanistic explanation of that disease. While interventionist accounts of causation and mechanistic explanations helped offer a fuller picture, Chapter 5 also highlighted that biomedical understanding is necessary but not sufficient for understanding in a clinical context (i.e., clinical understanding). The clinical context calls for combining *biomedical understanding* of a *disease* with *personal understanding* of an *illness.* This not only necessitates adopting the use of cognitive resources *in addition* to those involved in biomedical understanding, but also involves a type of understanding that cannot be reduced to grasping explanations.

Having clarified the epistemic aims of inquiry and the character of different types of understanding in medicine, we were in a position to say something about the final aim by detailing *how exactly* medicine contributes to supporting human agency. According to the *Autonomy Thesis*, medicine is not pathocentric, but aims to promote (positive) health with the final aim of enhancing autonomy. Chapter 6 offered a pluralist perspective on some problems with the concept "health" and defused the objection that the *Autonomy Thesis* is overly permissive. Completing the picture, Chapter 7 critically engaged contemporary accounts from the literature, focusing on examining to what extent they are able to overcome or bypass the challenges faced when defending the *Autonomy Thesis.* Subjecting these contemporary views to critical scrutiny is not merely an essentially adversarial procedure, but is also a means to assist framing the proposal presented in Chapter 6. By inspecting the most relevant aspects of these accounts in light of the challenges considered in Chapter 6, Chapter 7 also provided further reinforcement for the *Autonomy Thesis* by considering paths that the proposed account chose not to take.

C.3 The *Moderate Position* and the "Greatest Benefit to Mankind"

The final part of the book returned to the challenges to medicine, with the main task of expounding how the account defended in this book, along with the norms and values in medicine it underscored, can help address the three strands of criticism and the challenges to medicine they highlight. Taken together, the theses in this book support the *Moderate Position*, which, compared to more radical views, offers an improved comprehension of the relevant challenges, directs attention to possible solutions, and contributes to illuminating the appropriate boundaries of medicine. First, it was shown that increasing systematicity in simple and extended senses offers ways to address the challenge of skepticism, which helps explain why systematicity requires reconsidering resource allocation in medical research to prioritize certain research goals. Second, the *Moderate Position* identified two requirements that have to be fulfilled for the medicalization of a condition to be consistent with the aim of medicine, regardless of whether medicine can offer effective treatment for it. Third, the *Moderate Position* showed how objectification can hinder the pursuit of the aim of medicine by obstructing personal understanding and pointed to ways in which objectification can be counteracted.

These points begin to indicate why we need to free ourselves from the widely assumed narrative of scientific progress in medicine that inevitably leads to therapeutic successes. Medicine is a hybrid endeavor in which progress cannot merely consist in the creation of pharmacological, surgical, genetic, or other interventions, nor by the powers of the AI-driven technological revolution in which so many put their faith. Certainly, we may expect that innovations in AI, robotics, and big data will have a significant impact on our comprehension of disease. They will surely increase our biomedical understanding and abilities to predict and intervene, but it is hard to picture how they could solve challenges linked to personal understanding and the social disconnection that some health care environments promote and exhibit. The book has not provided definitive answers in these matters, but it has offered pointers for further research to ask new questions such as whether medical progress could extend to include clinical understanding and perhaps even something like access to care. In any case, denying neither the enormous potential these developments hold, nor the impressive advances already achieved, the *Moderate Position* stresses that medicine will fail to harvest the full benefits of these advances unless it implements measures that counteract the diminished personal understanding (and objectification) to which they can contribute. Only then will medicine exploit its full potential and move closer to living up to Samuel Johnson's famous accolade to medicine as "the greatest benefit to mankind."

It is often said that philosophy thrives in times characterized by disorientation and crisis. Perhaps this is not entirely true, but it is not difficult to see what might motivate such a conclusion. The type of disorientation that characterizes medicine's current circumstances both necessitates and provides a fertile ground for posing the type of fundamental questions with which philosophical work is well equipped to assist. Addressing them might prove useful beyond medicine, as aims pursued by other branches of science that acquire population-level data and biological knowledge of health and disease are not disconnected from the aim of medicine. It is hoped that this book has been able to clarify the relevance and benefits of philosophical reflection for debates on the future of medicine. The conclusions and solutions proposed should of course not be taken to suggest that the philosophical work in this area is somehow complete, but only that a productive course has been charted that merits further exploration. That said, the relevance of philosophical work in these matters should not make us blind to some important limitations that arise from the nature of philosophical work and from having to balance several aspirations. While acknowledging that no narrowly conceived nature and aim will capture the full complexity of medicine, this book tried to offer an account that is sufficiently broad to assist in addressing the challenges, yet sufficiently narrow to pass the requirements of philosophical meticulousness. The hope is that this book managed to strike the right balance and will inspire further work at the intersection of philosophy and medicine.

References

Alvarez, M. (2016). Reasons for action: Justification, motivation, explanation. *The Stanford Encyclopedia of Philosophy* (Winter 2017 Edition). Edward N. Zalta (ed.). Stanford University Press. https://plato.stanford.edu/archives/win2017/entries/reasons-just-vs-expl/

American College of Obstetricians and Gynecologists. (2017). Sterilization of women: Ethical issues and considerations. Committee Opinion No. 695. *Obstet Gynecol, 129*, e109–16.

Ananth, M. (2008). *In defense of an evolutionary concept of health: Nature, norms, and human biology*. Routledge.

Anderson, E. (1995). *Value in ethics and economics*. Harvard University Press.

Appiah, K. A. (1998). Race, culture, identity: Misunderstood connections. In Appiah, A., and Gutmann, A. (eds.). *Color conscious: The political morality of race* (pp. 30–105). Princeton University Press.

(2003). *Thinking it through: An introduction to contemporary philosophy*. Oxford University Press.

Aronowitz, R. (2019). What is medicine for? *Boston Review*. https://bostonreview.net/science-nature/robert-aronowitz-what-medicine

Aronson, J. K., and Green, A. R. (2020). Me-too pharmaceutical products: History, definitions, examples, and relevance to drug shortages and essential medicines lists. *Br J Clin Pharmacol, 86*, 2114–22.

Astin, J. A. (1998). Why patients use alternative medicine: Results of a national study. *JAMA, 279*(19), 1548–53.

Ayer, A. J. (1946). *Language, truth and logic*. Gollancz.

Baker, M. (2016). Is there a reproducibility crisis? *Nature, 533*, 452–4.

Ban, T. A. (2022). The role of serendipity in drug discovery. *Dialog Clin Neurosci, 8*(3), 335–44.

Barnes, P. M., Bloom, B., and Nahin, R. L. (2007). Complementary and alternative medicine use among adults and children: United States, 2007. National Health Statistics Reports, no 12. Hyattsville, MD: National Center for Health Statistics.

Baumberger, C. (2019). Explicating objectual understanding: Taking degrees seriously. *J Gen Philos Sci, 50*(3), 367–88.

Baumberger, C., Beisbart, C., and Brun, G. (2017). What is understanding? An overview of recent debates in epistemology and philosophy of science.

In Grimm, S., Baumberger, C., and Ammon, S. (eds.). *Explaining understanding: New perspectives from epistemology and philosophy of science* (pp. 1–34). Routledge.

Bayne, T., and Levy, N. (2005). Amputees by choice: Body integrity identity disorder and the ethics of amputation. *J Appl Philos, 22*(1), 758–6.

Beauchamp, T. L., and Childress, J. F. (2001). *Principles of biomedical ethics*. 5th ed. Oxford University Press.

Bechtel, W., and Abrahamsen, A. (2005). Explanation: A mechanist alternative. *Stud Hist Philos Biol Biomed Sci, 36*(2), 421–41.

Bellavite, P., Conforti, A., Piasere, V., and Ortolani, R. (2005). Immunology and homeopathy. 1. Historical background. *Evid Based Complement Alternat Med, 2*(4), 441–52.

Bengson, J. (2017). The unity of understanding. In Grimm, S. R. (ed.). *Making sense of the world: New essays on the philosophy of understanding* (pp. 14–53). Oxford University Press.

Ben-Moshe, N. (2019). The internal morality of medicine: A constructivist approach. *Synthese, 196*(11), 4449–67.

Berglund, M., Westin, L., Svanström, R., and Sundler, A. J. (2012). Suffering caused by care – Patients' experiences from hospital settings. *Int J Qual Stud Health Well-being, 7*, 1–9.

Berofsky, B. (1995). *Liberation from self: A theory of personal autonomy*. Cambridge University Press.

Bird, A. (2007). What is scientific progress? *Nous, 41*(1), 64–89.

(2019a). The aim of belief and the aim of science. *Theoria, 34*(2), 171–93.

(2019b). Systematicity, knowledge, and bias. How systematicity made clinical medicine a science. *Synthese, 196*(3), 863–79.

Bivins, R. E. (2010). *Alternative medicine? A history*. Oxford University Press.

Blumer, R., and Meyer, M. (2006). *The new medicine: Companion book to the public television series*. Middlemarch Films.

Boetto, E., Golinelli, D., Carullo, G., and Fantini, M. P. (2021). Frauds in scientific research and how to possibly overcome them. *J Med Ethics, 47* (12), e19.

Boorse, C. (1975). On the distinction between disease and illness. *Philos Public Aff, 5*, 49–68.

(1977). Health as a theoretical concept. *Philos Sci, 44*, 542–73.

(1997). A rebuttal on health. In Humber, J. M., and Almeder, R. F. (eds.). *What is disease?* (pp. 1–134). Humana Press.

(2014). A second rebuttal on health. *J Med Philos, 39*(6), 683–724.

(2016). Goals of medicine. In Giroux, É. (ed.). *Naturalism in the philosophy of health. History, philosophy and theory of the life sciences* (pp. 145–77). Springer International Publishing.

Brandom, R. (1994). *Making it explicit*. Harvard University Press.

Brandt, A. M., and Gardner, M. (2020). The golden age of medicine? In Cooter, A., and Pickstone, J. (eds.). *Medicine in the twentieth century* (pp. 21–37). Taylor & Francis.

Broadbent, A. (2011). Inferring causation in epidemiology: Mechanisms, black boxes, and contrasts. In McKay Illari, P., Russo, F., and Williamson, J. (eds.). *Causality in the sciences* (pp. 45–69). Oxford University Press.

(2019). *Philosophy of medicine*. Oxford University Press.

Brody, H., and Miller, F. G. (1998). The internal morality of medicine: Explication and application to managed care. *J Med Philos*, *23*(4), 384–410.

Broom, D. H., and Woodward, R. V. (1996). Medicalisation reconsidered: Toward a collaborative approach to care. *Sociol Health Illness*, *18*(3), 3573–8.

Brülde, B. (2001). The goals of medicine. Towards a unified theory. *Health Care Anal*, *9*(1), 1–13.

Burgess, A., and Plunkett, D. (2013). Conceptual ethics I. *Philos Compass*, *8*(12), 1091–101.

Bynum, B. (2008). The McKeown thesis. *Lancet*, *371*(9613), 644–5.

Callahan, D. (1996). The goals of medicine. Setting new priorities. *Hastings Center Report*, *26*(6), S1–27.

Caplan, A. L. (1992). Does the philosophy of medicine exist? *Theor Med*, *13*(1), 67–77.

Capozza, D. (2016). Dehumanization in medical contexts: An expanding research field. *TPM Test Psychom Methodol Appl Psychol*, *1*, 545–59.

Cappelen, H. (2018). *Fixing language: An essay on conceptual engineering*. Oxford University Press.

Carnap, R. (1950). *Logical foundations of probability*. University of Chicago Press.

Carpenter, K. J. (2012). The discovery of vitamin C. *Ann Nutr Metab*, *61*(3), 2592–64.

Carter, J. A., and Gordon, E. C. (2014). Objectual understanding and the value problem. *Am Philos Quart*, *51*(1), 1–13.

Cartwright, N. (1999). *The dappled world: A study of the boundaries of science*. Cambridge University Press.

Cassell, E. J. (2004). *The nature of suffering and the goals of medicine*. 2nd ed. Oxford University Press.

Cavell, S. (1969). Knowing and acknowledging. In *Must we mean what we say?* (Ch. IX, pp. 238–66). Cambridge University Press.

Centers for Disease Control and Prevention. (2020). Disability and Health Information for People with Disabilities. www.cdc.gov/ncbddd/disabilityandhealth/people.html

Chalmers, D. J. (2011). Verbal disputes. *Philos Rev*, *120*(4), 515–66.

(2015). Why isn't there more progress in philosophy? *Philosophy*, *90*(1), 3–31.

(2020). What is conceptual engineering and what should it be? *Inquiry*, DOI: 10.1080/0020174X.2020.1817141

Chambers, C. (2008). *Sex, culture, and justice: the limits of choice*. Penn State University Press.

Cheng, Y., Lin, C. P., Liu, H. L., et al. (2007). Expertise modulates the perception of pain in others. *Curr Biol*, *17*(19), 1708–13.

Christensen, D. (2010). Higher-order evidence. *Philos Phenom Res*, *81*(1), 185–215.

Christman, J. (2009). *The politics of persons: Individual autonomy and socio-historical selves*. Cambridge University Press.
Combs, Jr, G. F., and McClung, J. P. (2016). *The vitamins: Fundamental aspects in nutrition and health*. Academic Press.
Conrad, P. (2007). *The medicalization of society: On the transformation of human conditions into treatable disorders*. Johns Hopkins University Press.
Conrad, P., Mackie, T., and Mehrotra, A. (2010). Estimating the costs of medicalization. *Soc Sci Med*, *70*(12), 1943–7.
Coriddi, M., Nadeau, M., Taghizadeh, M., and Taylor, A. (2013). Analysis of satisfaction and well-being following breast reduction using a validated survey instrument: The BREAST-Q. *Plast Reconstr Surg*, *132*(2), 285–90.
Coulehan, J. L., Platt, F. W., Egener, B., et al. (2001). Let me see if i have this right. . .": Words that help build empathy. *Ann Intern Med*, *135*(3), 221–7.
Craver, C. F. (2007). *Explaining the brain: Mechanisms and the mosaic unity of neuroscience*. Oxford University Press.
Craver, C. F., and Bechtel, W. (2007). Top-down causation without top-down causes. *Biol Philos*, *22*(4), 547–63.
Craver, C. F., and Darden, L. (2013). *In search of mechanisms: Discoveries across the life sciences*. University of Chicago Press.
Craver, C. F., and Tabery, J. (2019). Mechanisms in science. *The Stanford Encyclopedia of Philosophy* (Summer 2019 Edition). https://plato.stanford.edu/archives/sum2019/entries/science-mechanisms
Craver, C. F., Glennan, S., and Povich, M. (2021). Constitutive relevance & mutual manipulability revisited. *Synthese*, *199*(3), 8807–28.
Cunningham, T. V. (2015). Objectivity, scientificity, and the dualist epistemology of medicine. In Huneman, P., Lambert, G., and Silberstein, M. (eds.). *Classification, disease and evidence* (Vol. 7, pp. 1–17). Springer.
Cutler, D. M. (2004) *Your money or your life*. Oxford University Press.
Cutler, D. M., Rosen, A. B., and Vijan, S. (2006a). The value of medical spending in the United States, 1960–2000. *N Engl J Med*, *355*(9), 920–7.
Cutler, D., Deaton, A., and Lleras-Muney, A. (2006b). The determinants of mortality. *JEP*, *20*(3), 97–120.
Daly, P. (2017). Philosophy of medicine 2017: Reviewing the situation. *Theor Med Bioeth*, *38*(6), 483–8.
Danish Health and Medicines Authority. (2014), The seven roles of physicians. www.sst.dk/en/news/2013/-/media/39D3E216BCBF4A9096B286EE44F03691.ashx
Darrason, M. (2018). Mechanistic and topological explanations in medicine: The case of medical genetics and network medicine. *Synthese*, *195*(1), 147–73.
Davidson, D. (1963). Actions, reasons, and causes. In Davidson, D. (ed.) *Essays on actions and events* (pp. 3–20). Clarendon Press.
Davis, K. (1991). Remaking the she-devil: A critical look at feminist approaches to beauty. *Hypatia*, *6*(2), 21–43.
Dawson, J. (2021). Medically optimised: Healthcare language and dehumanisation. *Br J Gen Pract*, *71*(706), 224.

Debes, R. (2018). Understanding persons and the problem of power. In Grimm, S. (ed.). *Making sense of the world: New essays on the philosophy of understanding* (pp. 547–7). Oxford University Press.

De Block, A., and Hens, K. (2021). A plea for an experimental philosophy of medicine. *Theor Med Bioeth*, *42*(3), 81–9.

Dellsén, F. (2016). Scientific progress: Knowledge versus understanding. *Stud Hist Philos Sci Part A*, *56*, 72–83.

Dennett, D. (1986). Philosophy as mathematics or as anthropology. *Mind Language* *1*(1), 18–19.

De Regt, H. W. (2009). The epistemic value of understanding. *Philos Sci*, *76*(5), 585–97.

(2017). *Understanding scientific understanding*. Oxford University Press.

De Regt, H. W., and Dieks, D. (2005). A contextual approach to scientific understanding. *Synthese*, *144*(1), 137–70.

De Regt, H. W., Leonelli, S., and Eigner, K. (2009). Focusing on scientific understanding. In De Regt, H. W., Leonelli, S., and Eigner, K. (eds.). *Scientific understanding: Philosophical perspectives* (pp. 11–17). University of Pittsburgh Press.

De Vreese, L. (2008). Causal (mis) understanding and the search for scientific explanations: a case study from the history of medicine. *Stud Hist Philos Sci Part C*, *39*(1), 14–24.

Dewalt, D. A., and Pincus, T. (2003). The legacies of Rudolf Virchow: Cellular medicine in the 20th century and social medicine in the 21st century. *Isr Med Assoc J*, *5*(6), 395–7.

Di Bernardo, G. A., Visintin, E. P., Dazzi, C., and Capozza, D. (2011). Patients' dehumanization in health contexts. *Poster presented at the 12th Annual Meeting of the Society for Personality and Social Psychology*, San Antonio, TX.

Douglas, H. (2003). The moral responsibilities of scientists (tensions between autonomy and responsibility). *Am Philos Quart*, *40*(1), 59–68.

(2009). *Science, policy, and the value-free ideal*. University of Pittsburgh Press.

Drake, J. M., Brett, T. S., Chen, S., et al. (2019). The statistics of epidemic transitions. *PLoS Comput Biol*, *15*(5), e1006917.

Dummett, M. (2010). *The nature and future of philosophy*. Columbia University Press.

Dupré, J. (1993). *The disorder of things: Metaphysical foundations of the disunity of science*. Harvard University Press.

(2004). The miracle of monism. In de Caro, M., and MacArthur, D. (eds.). *Naturalism in question* (pp. 21–39). Harvard University Press.

(2016). Toward a political philosophy of science. In Couch, M., and Pfeifer, J. (eds.). *The philosophy of Philip Kitcher* (pp. 182–205). Oxford University Press.

Eklund, M. (2014). Replacing truth. In Burgess, A., and Sherman, B. (eds.). *Metasemantics: New essays on the foundations of meaning* (pp. 293–310). Oxford University Press.

(2015). Intuitions, conceptual engineering, and conceptual fixed points. In Daly, C. (ed.). *The Palgrave handbook of philosophical methods* (pp. 363–85). Palgrave Macmillan.
Elgin, C. (2007). Understanding and the facts. *Philos Stud, 132*(1), 33–42.
(2009a). Is understanding factive? In Haddock, A., Millar, A., and Pritchard, D. (eds.). *Epistemic value* (pp. 322–30). Oxford University Press.
(2009b). Exemplification, idealization, and understanding. In Suárez, M. (ed.). *Fictions in science: Essays on idealization and modelling* (pp. 77–90). Routledge.
Elgin, C. Z. (2017). *True enough*. MIT Press.
Engelhardt, Jr, H. T., and Jotterand, F. (eds.). (2008). *The philosophy of medicine reborn: A Pellegrino reader*. University of Notre Dame Press.
Ereshefsky, M. (2009). Defining "health" and "disease." *Stud Hist Philos Sci Part C, 40*(3), 221–7.
Ernst, E., and Singh, S. (2008). *Trick or treatment: The undeniable facts about alternative medicine*. W. W. Norton.
Ezell, J. M. (2022). The medicalization of freedom: How anti-science movements use the language of personal liberty and how we can address it. *Nat Med, 28* (2), 219.
Fanelli, D. (2018). Is science really facing a reproducibility crisis? *Proc Natl Acad Sci U S A, 115*(11), 2628–31.
Farmer, P., Basilico, M., and Messac, L. (2016). After McKeown. In Greene, J., Condrau, F., and Watkins, E. S. (eds.). *Therapeutic revolutions* (pp. 186–217). University of Chicago Press.
Flory, J. H., and Kitcher, P. (2004). Global health and the scientific research agenda. *Philos Public Aff, 32*(1), 36–65.
Fuchs, V. R. (2010). New priorities for future biomedical innovations. *N Engl J Med, 363*(8), 704–6.
Gadamer, H. G. (1996). *The enigma of health* (trans: Gaiger, J., and Walker, N.). Polity.
Gallagher, S. (2005). *How the body shapes the mind*. Clarendon Press.
Galvani, A. P., Parpia, A. S., Pandey, A., et al. (2022). Universal healthcare as pandemic preparedness: The lives and costs that could have been saved during the COVID-19 pandemic. *Proc Natl Acad Sci U S A, 119*(25), e2200536119.
Gijsbers, V. (2013). Understanding, explanation, and unification. *Stud Hist Philos Sci Part A, 44*(3), 516–22.
Gillett, C. (2020). Why constitutive mechanistic explanation cannot be causal: Highlighting needed theoretical projects and their constraints. *Am Philos Quart, 57*(1), 31–50.
Glennan, S. (2017). *The new mechanical philosophy*. Oxford University Press.
Glennan, S., Illari, P., and Weber, E. (2022). Six theses on mechanisms and mechanistic science. *J Gen Philos Sci, 53*(2), 143–61.
Goldman, A. (2006). *Simulating minds: The philosophy, psychology, and neuroscience of mindreading*. Oxford University Press.

Goodman, S., and Greenland, S. (2007). Why most published research findings are false: Problems in the analysis. *PLoS Med*, *4*(4), e168.

Gopnik, A. (1998). Explanation as orgasm. *Minds Mach, 8*, 101–18.

(2000). Explanation as orgasm and the drive for causal knowledge: The function, evolution, and phenomenology of the theory formation system. In Keil, F. C., and Wilson, R. A. (eds.). *Explanation and cognition* (pp. 299–323). MIT Press.

Grimm, S. R. (2006). Is understanding a species of knowledge? *Br J Philos Sci, 57*(3).

(2011). Understanding. In Bernecker, S., and Pritchard, D. (eds.). *The Routledge companion to epistemology* (pp. 84–94). Routledge.

(2012). The value of understanding. *Philos Compass*, *7*(2), 103–17.

(2014). Understanding as knowledge of causes. In Fairweather, A. (ed.). *Virtue epistemology naturalized: Bridges between virtue epistemology and philosophy of science* (pp. 329–45). Springer.

(2016). How understanding people differs from understanding the natural world. *Philos Issues, 26*(1), 209–25.

(2017). Understanding and transparency. In Grimm, S., Bamberger, C., and Ammon, S. (eds.). *Explaining understanding: New perspectives from epistemology and philosophy of science* (pp. 212–29). Routledge.

(2021). Understanding. *The Stanford Encyclopedia of Philosophy* (Summer 2021 Edition), Edward N. Zalta (ed.), https://plato.stanford.edu/archives/sum2021/entries/understanding/

Halpern, J. (2014). From idealized clinical empathy to empathic communication in medical care. *Med Health Care Philos 17*, 301–11.

Hannon, M. (2019). *What's the point of knowledge?: A function-first epistemology*. Oxford University Press.

(2021). Recent work in the epistemology of understanding. *Am Philos Quart, 58*(3), 269–90.

Hansson, S. O. (2017). Science and pseudo-science. *The Stanford Encyclopedia of Philosophy* (Summer 2017 Edition). Edward N. Zalta (ed.). https://plato.stanford.edu/archives/sum2017/entries/pseudo-science/

Haque, O. S., and Waytz, A. (2012). Dehumanization in medicine: Causes, solutions, and functions. *Perspect Psychol Sci*, *7*(2), 176–86.

Haslam, N., Loughnan, S., Reynolds, C., and Wilson, S. (2007). Dehumanization: A new perspective. *Soc Personal Psychol Compass, 1*(1), 409–22.

Haslanger, S. (2000). Gender and race: (What) are they? (What) do we want them to be? *Noûs, 34*(1), 31–55.

(2018). What is a social practice? *R Inst Philos Suppl, 82*, 231–47.

Hausman, D. (2017). Health and well-being. In Solomon, M., Simon, J., and Kincaid, H. (eds.). *The Routledge companion to philosophy of medicine* (pp. 273–5). Routledge.

Hawking, S., and Mlodinow, L. (2010). *The grand design*. Bantam Dell Publishing Group.

Heal, J. (2003). *Mind, reason and imagination*. Cambridge University Press.

Hempel, C. G. (1966). *Philosophy of natural science*. Prentice-Hall.
Hempel, C. G., and Oppenheim, P. (1948). Studies in the logic of explanation. *Philos Sci, 15*(2), 135–75.
Henderson, D. K., and Horgan, T. (2011). *The epistemological spectrum: At the interface of cognitive science and conceptual analysis*. Oxford University Press.
Hershenov, D. B. (2020). Pathocentric health care and a minimal internal morality of medicine. *J Med Philos, 45*(1), 16–27.
Hills, A. (2016). Understanding why. *Noûs, 50*(4), 661–88.
Hofmann, B. (2016). Disease, illness, and sickness. In Solomon, M., Simon, J., and Kincaid, H. (eds.). *The Routledge companion to philosophy of medicine* (pp. 30–40). Routledge.
Hojat, M., Louis, D. Z., Markham, F. W., Wender, R., Rabinowitz, C., and Gonnella, J. S. (2011). Physicians' empathy and clinical outcomes for diabetic patients. *Acad Med, 86*, 359–64.
Honneth, A. (2008). *Reification: A new look at an old idea*. Oxford University Press.
Horton, R. (2015). Offline: What is medicine's 5 sigma. *Lancet, 385*(9976), 1380.
Howick, J. (2011). Exposing the vanities – and a qualified defence – of mechanistic evidence in clinical decision-making. *Philos Sci, 78*, 926–40.
Howick, J., Mittoo, S., Abel, L., Halpern, J., and Mercer, S. W. (2020). A price tag on clinical empathy? Factors influencing its cost-effectiveness. *J R Soc Med, 113*(10), 389–93.
Hoyningen-Huene, P. (2013). *Systematicity: The nature of science*. Oxford University Press.
(2019). Replies. *Synthese, 196*(3), 907–28.
Huber, M., Knottnerus, J. A., Green, L., et al. (2011). How should we define health? *BMJ, 343*.
Huneman, P., Lambert, G., and Silberstein, M. (2015). Introduction: Surveying the revival in the philosophy of medicine. In Huneman, P., Lambert, G., and Silberstein, M. (eds.). *Classification, disease and evidence: New essays in the philosophy of medicine* (pp. vii–xix). Springer.
Illari, P. M. (2011). Disambiguating the Russo-Williamson thesis. *Int Stud Philos Sci, 25*, 139–57.
Illari, P. M., and Williamson, J. (2012). What is a mechanism? Thinking about mechanisms across the sciences. *Eur J Philos Sci, 2*(1), 119–35.
Illich, I. (1974). Medical nemesis. *Lancet, 303*(7863), 918–21.
Ioannidis, J. P. (2005). Why most published research findings are false. *PLoS Med, 2*(8), e124.
(2016). Why most clinical research is not useful. *PLoS Med, 13*(6), e1002049.
Jaeggi, R. (2018). *On the critique of forms of life*. Belknap Press.
Jamshidi, S., Parker, J. S., and Hashemi, S. (2020). The effects of environmental factors on the patient outcomes in hospital environments: A review of literature. *Front Archit Res*. https://doi.org/10.1016/j.foar.2019.10.001
Johnson, M. (2015). Embodied understanding. *Front Psychol, 6*, 875.

Kaiser, M. I. (2019). Normativity in the philosophy of science. *Metaphilosophy, 50* (1–2), 36–62.
Kaplan, D. M., and Craver, C. F. (2011). The explanatory force of dynamical and mathematical models in neuroscience: A mechanistic perspective. *Philos Sci, 78*(4), 601–27.
Kaplan, R. M., and Milstein, A. (2019). Contributions of health care to longevity: A review of 4 estimation methods. *Ann Fam Med, 17*(3), 267–72.
Kauppinen, A. (2002). Reason, recognition, and internal critique. *Inquiry, 45*(4), 479–98.
(2007). The rise and fall of experimental philosophy. *Philos Explor, 10*(2), 95–118.
Kelly, T. (2016). Evidence. *The Stanford Encyclopedia of Philosophy.* Edward N. Zalta (ed.). https://plato.stanford.edu/archives/win2016/entries/evidence/
Kelp, C. (2021). Inquiry, knowledge and understanding. *Synthese, 198*(7), 1583–93.
Kendler, K. S., and Campbell, J. (2009). Interventionist causal models in psychiatry: Repositioning the mind–body problem. *Psychol Med, 39*(6), 881–7.
Kendler, K. S., and Neale, M. C. (2010). Endophenotype: A conceptual analysis. *Mol Psychiatry, 15*(8), 789–97.
Kernahan, P. J. (2012) Was there ever a 'golden age' of medicine?. *Minn Med, 95*(9), 41–5.
Kersey, R. D., Elliot, D. L., Goldberg, L., et al. (2012). National Athletic Trainers' Association position statement: Anabolic-androgenic steroids. *J Athl Train, 47*(5), 567–88.
Khalifa, K. (2013). The role of explanation in understanding. *Br J Philos Sci, 64*(1), 161–87.
(2017). *Understanding, explanation, and scientific knowledge.* Cambridge University Press.
(2019). Is *verstehen* scientific understanding? *Philos Soc Sci, 49*(4), 282–306.
Kingma, E. M. (2012). Health and health promotion. Technische Universiteit Eindhoven. https://pure.tue.nl/ws/files/3462442/kingma2012.pdf
Kingma, E. (2019). Contemporary accounts of health. In Kingma, E. (ed.). *Health* (pp. 289–318). Oxford University Press.
Kinzer, S. (2019). *Poisoner in chief: Sidney Gottlieb and the CIA search for mind control.* Henry Holt and Company.
Kirmayer, L. J. (2011). Multicultural medicine and the politics of recognition. *J Med Philos, 36*(4), 410–23.
Kirsch, I. (2019). Placebo effect in the treatment of depression and anxiety. *Front Psychiatry, 10*, 407.
Kitcher, P. (2001). *Science, truth, and democracy.* Oxford University Press.
(2008). Carnap and the caterpillar. *Philos Top, 36*(1), 111–27.
(2011). *Science in a democratic society.* Prometheus Books.
(2015). Pragmatism and progress. *Trans Charles S. Peirce Soc, 51*(4), 475.

(2016). Reply to Dupré. In Couch, M. (ed.). *The philosophy of Philip Kitcher* (pp. 182–205). Oxford University Press.

Kornblith, H. (1993). Epistemic normativity. *Synthese*, *94*(3), 357–76.

(2002). *Knowledge and its place in nature*. Oxford University Press.

(2017). A naturalistic methodology. In D'Oro, G., and Overgaard, S. (eds.). *The Cambridge companion to philosophical methodology* (1st ed., pp. 141–60). Cambridge University Press.

Kraut, R. (2009). *What is good and why: The ethics of well-being*. Harvard University Press.

Kripke, S. (1980). *Naming and necessity*. Harvard University Press.

Kuhn, T. S. (1962). *The structure of scientific revolutions*. University of Chicago Press.

(1974). Logic of discovery or psychology of research? In Schilpp, P. A. (ed.). *The Philosophy of Karl Popper*, The Library of Living Philosophers (Vol. xiv, book ii, pp. 798–819). Open Court.

(1977). *The essential tension. Selected studies in scientific tradition and change*. University of Chicago Press.

Kukla, R. (2005). Conscientious autonomy: Displacing decisions in health care. *Hastings Center Report*, *35*(2), 344–4.

Kvanvig, J. L. (2003). *The value of knowledge and the pursuit of understanding*. Cambridge University Press.

Kvanvig, J. (2009). The value of understanding. In Haddock, A., Millar, A., and Pritchard, D. (eds.). *Epistemic value* (pp. 95–111). Oxford University Press.

(2013). Curiosity and the response-dependent special value of understanding. In Henning, T., and Schweikard, D. P. (eds.). *Knowledge, virtue and action: Putting epistemic virtues to work* (pp. 151–74). Routledge.

Kyle, R. A., Steensma, D. P., and Shampo, M. A. (2016). Barry James Marshall – Discovery of Helicobacter pylori as a cause of peptic ulcer. *Mayo Clin Proc*, *91*(5), e67–8.

Lakatos, I. (1981). Science and pseudoscience. In Brown, S., Fauvel, J., and Finnegan, R. (eds.). *Conceptions of inquiry: A reader* (pp. 114–21). Methuen.

Laplane, L., Mantovani, P., Adolphs, R., et al. (2019). Opinion: Why science needs philosophy. *Proc Natl Acad Sci U S A*, *116*(10), 3948–52.

Laudan, L. (1983). The demise of the demarcation problem. In Cohen, R. S., and Laudan, L. (eds.). *Physics, philosophy and psychoanalysis* (pp. 111–27). Springer.

(1984). *Science and values*. University of California Press.

Le Fanu, J. (2012). *The rise and fall of modern medicine*. Basic Books.

Lemoine, M. (2015). The naturalization of the concept of disease. In Hunemann, P., Lambert, G., and Silberstein, M. (eds.). *Classification, disease and evidence* (pp. 19–41). Springer.

Lemoine, M., Darrason, M., and Richard, H. (2014). Where is philosophy of medicine headed? A report of the International Advanced Seminar in the Philosophy of Medicine (IASPM). *J Eval Clin Pract*, *20*(6), 991–3.

Lewis, D. (1986). Causal explanation. In Lewis, D. (ed.). *Philosophical papers* (Vol. II, pp. 214–40). Oxford University Press.

Lichstein, P. R. (1990). The medical interview. In Walker, H. K., Hall, W. D., and Hurst, J. W. (eds.). *Clinical methods: The history, physical, and laboratory examinations* (3rd ed., Ch. 3, pp. 29–36). Butterworths.

Lind, J. (1772). *A treatise on the scurvy: In three parts. Containing an inquiry into the nature, causes, and cure, of that disease. Together with a critical and chronological view of what has been published on the subject.* 3rd ed. S. Crowder.

Lipton, P. (2001). What good is an explanation? In Hon, G., and Rakover, S. S. (eds.). *Explanation: Theoretical approaches and applications* (pp. 53–9). Springer Science and Business Media.

(2004). *Inference to the best explanation.* Routledge.

(2009). Understanding without explanation. In de Regt, H. W., Leonelli, S., and Eigner, K. (eds.). *Scientific understanding: Philosophical perspectives* (pp. 43–63). University of Pittsburgh Press.

Loeb, S., Bjurlin, M. A., Nicholson, J., et al. (2014). Overdiagnosis and overtreatment of prostate cancer. *Eur Urol, 65*(6), 1046–55.

Loomis, E., and Juhl, C. (2006). Explication. In Sarkar, S., and Pfeifer, J. (eds.). *The philosophy of science: An introduction* (pp. 287–94). Routledge.

Lyerly, A. D. (2012). Ethics and "normal birth." *Birth, 39*(4), 315–17.

Lyre, H. (2018). Medizin als Wissenschaft – eine wissenschaftstheoretische Analyse. In Ringkamp D., and Wittwer, H. (eds.). *Was ist Medizin? Der Begriff der Medizin und seine ethischenImplikationen* (pp. 143–66). Alber.

Machamer, P., Darden, L., and Craver, C. F. (2000). Thinking about mechanisms. *Philos Sci, 67*(1), 12–15.

MacIntyre, A. (2007). *After virtue: A study in moral theory [1981].* Duckworth.

Makary, M. A., and Daniel, M. (2016). Medical error – the third leading cause of death in the US. *Bmj, 353.*

Malacrida, C., and Boulton, T. (2014). The best laid plans? Women's choices, expectations and experiences in childbirth. *Health, 18*(1), 41–59.

Marcum, J. A. (2008). *Humanizing modern medicine: An introductory philosophy of medicine.* Springer.

(2012). Medicine's crises. In Marcum, J. A. (ed.). *The virtuous physician* (Vol. 114, pp. 1–28). Springer.

(2016) *Bloomsbury companion to contemporary philosophy of medicine.* Bloomsbury.

Mardis, E. R., and Wilson, R. K. (2009). Cancer genome sequencing: A review. *Hum Mol Genet, 18*(R2), R163–8.

McAndrew, S. (2019). Internal morality of medicine and physician autonomy. *J Med Ethics, 45*(3), 198–203.

McCrory, C., and McNally, S. (2013). The effect of pregnancy intention on maternal prenatal behaviours and parent and child health: Results of an Irish cohort study. *Paediatr Perinat Epidemiol, 27*(2), 208–15.

McGinn, C. (2015). The science of philosophy. *Metaphilosophy, 46*(1), 84–103.

McKenna, M. (2020). The antibiotic paradox: Why companies can't afford to create life-saving drugs. *Nature*, *584*(7821), 338–41.
McKeown, T. (1976). *The modern rise of population*. Academic Press.
McMullin, E. (1983). Values in science. In Asquith, P. D., and Nickles, T. (eds.). *Proceedings of the 1982 Biennial Meeting of the Philosophy of Science Association* (Vol. 1, pp. 32–8). Philosophy of Science Association.
Messerli, F. H. (2012). Chocolate consumption, cognitive function, and Nobel laureates. *N Engl J Med*, *376*, 1562–4.
Miller, C. (2014). Medicine is not science: Guessing the future, predicting the past: Towards a theory of irregularity. *J Eval Clin Pract*, *20*(6), 865–71.
Miller, F. G., and Brody, H. (2001). The internal morality of medicine: An evolutionary perspective. *J Med Philos*, *26*(6), 581–99.
Miller, C., and Miller, D. (2014). Medicine is not science. *EJPCH*, *2*(2), 144–53.
Miller, F. G., Brody, H., and Chung, K. C. (2000). Cosmetic surgery and the internal morality of medicine. *Camb Q Healthc Ethics*, *9*, 353.
Mintzes, B. (2006). Disease mongering in drug promotion: Do governments have a regulatory role? *PLoS Med*, *3*(4), e198.
Morganti, M. (2016). Naturalism and realism in the philosophy science. In Clark, K. J. (ed.). *The Blackwell companion to naturalism* (pp. 75–90). John Wiley & Sons.
Moynihan, R., and Cassels, A. (2005). *Selling sickness: How drug companies are turning us all into patients*. Crowsnest.
Mukerji, N., and Ernst, E. (2022). Why homoeopathy is pseudoscience. *Synthese*, *200*(5), 1–29.
Munson, R. (1981). Why medicine cannot be a science. *J Med Philos*, *6*(2), 183–208.
Murphy, D. (2005). Can evolution explain insanity? *Biol Philos*, *20*(4), 745–66.
Nagel, E. (1961). *The structure of science: Problems in the logic of scientific explanation*. Harcourt, Brace, and World, Inc.
Nervi, M. (2010). Mechanisms, malfunctions and explanation in medicine. *Biol Philos*, *25*, 215–28.
Neumann, M., Bensing, J., Mercer, S., Ernstmann, N., Ommen, O., and Pfaff, H. (2009). Analyzing the "nature" and "specific effectiveness" of clinical empathy: A theoretical overview and contribution towards a theory-based research agenda. *Patient Educ Couns*, *74*, 339–46.
Newman, M. (2012). An inferential model of scientific understanding. *Int Stud Philos Sci*, *26*(1), 1–26.
Ng, C. C. W., and Saad, T. C. (2021). Re-examining the idea of internal morality in medicine. *New Bioeth*, *27*(3), 230–44.
Nichols, S. (2002). How psychopaths threaten moral rationalism: Is it irrational to be amoral? *Monist*, *85*(2), 285–303.
Nissen, S. E. (2010). The rise and fall of rosiglitazone. *Eur Heart J*, *31*(7), 773–6.
Nordenfelt, L. (1987). *On the nature of health: An action-theoretic approach* (Vol. 26). Springer.

(1998). On medicine and health enhancement – towards a conceptual framework. *Med Health Care Philos, 1*(1), 51–2.
(2007). The concepts of health and illness revisited. *Med Health Care Philos, 10* (1), 5–10.
(2017). On concepts of positive health. In Schramme, T., and Edwards, S. (eds.). *Handbook of the philosophy of medicine* (pp. 29–43). Springer.
O'Mahony, S. (2019a). After the golden age: What is medicine for? *Lancet, 393* (10183), 1798–9.
(2019b). *Can medicine be cured?: The corruption of a profession.* Head of Zeus Ltd.
Oreskes, N. (2019). Systematicity is necessary but not sufficient: On the problem of facsimile science. *Synthese, 196*(3), 881–905.
Parens, E. (2013). On good and bad forms of medicalization. *Bioethics, 27*(1), 28–35.
Pellegrino, E. D. (1976). Philosophy of medicine: Problematic and potential. *J Med Philos, 1*(1), 53–1.
(1986). Philosophy of medicine: Towards a definition. *J Med Philos, 11*(1), 9–16.
(1998). What the philosophy of medicine is. *Theor Med Bioeth, 19*(4), 315–36.
(2001). The internal morality of clinical medicine: A paradigm for the ethics of the helping and healing professions. *J Med Philos, 26*(6), 559–79.
Pellegrino, E. D., and Thomasma, D. C. (1981). *A philosophical basis of medical practice.* Oxford University Press.
(1993). *The virtues in medical practice.* Oxford University Press.
Pennock, R. T. (2019). *An instinct for truth: Curiosity and the moral character of science.* MIT Press.
Pigliucci, M. (2015). Scientism and pseudoscience: A philosophical commentary. *J Bioeth Inq, 12*, 569–75.
Pigliucci, M., and Boudry, M. (2013). Introduction: Why the demarcation problem matters. In Pigliucci, M., and Boudry, M. (eds.). *Philosophy of pseudoscience: Reconsidering the demarcation problem* (pp. 1–16). The University of Chicago Press.
Popper, K. (1962). *Conjectures and refutations. The growth of scientific knowledge.* Basic Books.
Popper, K. R. (1979). *Objective knowledge: An evolutionary approach.* Oxford University Press.
(2000). *Realism and the aim of science: From the postscript to the logic of scientific discovery.* Routledge.
Porter, R. (1997/1999). *The greatest benefit to mankind: A medical history of mankind.* WW Norton and Company.
(2002). *Blood and guts: A short history of medicine.* W. W. Norton.
Potochnik, A. (2015). The diverse aims of science. *Stud Hist Philos Sci Part A, 53*, 71–80.
(2017). *Idealization and the aims of science.* University of Chicago Press.

Prinz, J. (2007). *The emotional construction of morals*. Oxford University Press.
Pritchard, D. (2010). Knowledge and understanding. In *The nature and value of knowledge: Three investigations*, Part I. Co-authored with Millar, A., and Haddock, A. Oxford University Press.
(2016). Seeing it for oneself: Perceptual knowledge, understanding, and intellectual autonomy. *Episteme*, *13*(1), 29–42.
Radiological Society of North America (2008) Patient photos spur radiologist empathy and eye for detail. *ScienceDaily*, 14 Dec. www.sciencedaily.com/releases/2008/12/081202080809.htm
Raffaeli, W., and Arnaudo, E. (2017). Pain as a disease: An overview. *J Pain Res*, *10*, 2003–8.
Rakel, D., Barrett, B., Zhang, Z., et al. (2011). Perception of empathy in the therapeutic encounter: Effects on the common cold. *Patient Educ Couns*, *85*, 390–7.
Ramsey, W. (1992). Prototypes and conceptual analysis. *Topoi*, *11*(1), 59–70.
Rawls, J. (1971). *A theory of justice*. Oxford University Press.
Redberg, R. F., and Katz, M. H. (2016). Statins for primary prevention: The debate is intense, but the data are weak. *JAMA*, *316*(19), 1979–81.
Reisch, G. A. (1998). Pluralism, logical empiricism, and the problem of pseudoscience. *Philos Sci*, *65*(2), 333–48.
Reiser, D. E., and Schroder, A. K. (1980). *Patient interviewing – The human dimension*. Williams and Wilkins.
Reiss, J., and Kitcher, P. (2009). Biomedical research, neglected diseases, and well-ordered science. *Theoria*, *24*(3), 263–82.
Riggs, W. D. (2003). Understanding "virtue" and the virtue of understanding. In DePaul, M. R., and Zagzebski, L. T. (eds.). *Intellectual virtue: Perspectives from ethics and epistemology* (pp. 203–26). Clarendon Press.
Risjord, M. (2014). *Philosophy of social science: A contemporary introduction*. Routledge.
Rosenberg, A. (2005). *Philosophy of science: A contemporary introduction*. Routledge.
Rowbottom, D. P. (2014). Aimless science. *Synthese*, *191*(6), 1211–21.
Ruse, M. (2009). *Monad to man: The concept of progress in evolutionary biology*. Harvard University Press.
Russo, F., and Williamson, J. (2007). Interpreting causality in the health sciences. *Int Stud Philos Sci*, *21*(2), 157–70.
Ryan, J., Khanda, G. E., Hibbert, R., et al. (2015). Is a picture worth a thousand words? The effect of viewing patient photographs on radiologist interpretation of CT studies. *J Am Coll Radiol*, *12*(1), 104–7.
Sadegh-Zadeh, K. (2012). *Handbook of analytic philosophy of medicine* (Vol. 113). Springer.
Salmon, W. C. (1984). *Scientific explanation and the causal structure of the world*. Princeton University Press.
Sandman, L., and Hansson, E. (2020). An ethics analysis of the rationale for publicly funded plastic surgery. *BMC Medical Ethics*, *21*(1), 1–14.

Scharp, K. (2013). *Replacing truth*. Oxford University Press.

Schramme, T. (2017a). Goals of medicine. In Schramme, T., and Edwards, S. (eds.). *Handbook of the philosophy of medicine* (pp. 121–8). Springer.

(2017b). Philosophy of medicine and bioethics. In Schramme, T., and Edwards, S. (eds.). *Handbook of the philosophy of medicine* (pp. 3–15). Springer.

Schramme, T., and Edwards, S. (eds.). (2017). *Handbook of the philosophy of medicine*. Springer.

Schroeder, S. A. (2013). Rethinking health: Healthy or healthier than? *Br J Philos Sci*, *64*(1), 131–59.

(2016). Health, disability, and well-being. In Fletcher, G. (ed.). *Routledge handbook of philosophy of well-being* (pp. 221–32). Routledge.

Schroer, J. W., and Schroer, R. (2013). Two potential problems with philosophical intuitions: Muddled intuitions and biased intuitions. *Philosophia*, *41*(4), 1263–81.

Schupbach, J. N. (2015). Experimental explication. *Philos Phenomenol Res*, *91*(2), 672–710.

Scott, S. (2006). The medicalisation of shyness: From social misfits to social fitness. *Sociol Health Illn*, *28*(2), 133–53.

Shojania, K. G., and Dixon-Woods, M. (2017). Estimating deaths due to medical error: The ongoing controversy and why it matters. *BMJ Qual Saf*, *26*(5), 423–8.

Sholl, J. (2017). The muddle of medicalization: Pathologizing or medicalizing?. *Theor Med Bioeth*, *38*(4), 2652–78.

(2021). Can aging research generate a theory of health? *Hist Philos Life Sci*, *43*(2), 45.

Sholl, J., and Rattan, S. I. (2020). How is "Health" explained across the sciences? Conclusions and recapitulation. In Sholl, J., and Rattan, S. I. (eds.). *Explaining health across the sciences* (pp. 541–9). Springer.

Smith, K. (2012). Against homeopathy – A utilitarian perspective. *Bioethics*, *26* (8), 398–409.

Snow, H. A., and Fleming, B. R. (2014). Consent, capacity and the right to say no. *Med J Aust*, *201*(8), 486–8.

Sober, E. (2008). *Evidence and evolution: The logic behind the science*. Cambridge University Press.

Solomon, M. (2015). *Making medical knowledge*. Oxford University Press.

Solomon, M., Simon, J. R., and Kincaid, H. (eds.). (2017). *The Routledge companion to philosophy of medicine*. Routledge, Taylor and Francis Group.

Sosa, E. (2007) Experimental philosophy and philosophical intuition. *Philos Stud*, *132*, 99–120.

Steel, D. (2008). *Across the boundaries, extrapolation in biology and social science*. Oxford University Press.

Stegenga, J. (2018). *Medical nihilism*. Oxford University Press.

Steinberger, F. (2016). The normative status of logic. *The Stanford Encyclopedia of Philosophy* (Spring 2017 Edition). Edward N. Zalta (ed.). https://plato.stanford.edu/archives/spr2017/entries/logic-normative/

Strevens, M. (2010). Varieties of understanding. In Pacific Division meeting of the American Philosophical Association, San Francisco, CA, March (Vol. 31).

(2013). No understanding without explanation. *Stud Hist Philos Sci Part A*, *44*(3), 510–15.

(2017). How idealizations provide understanding. In Grimm, S. R. (ed.). *Explaining understanding: New perspectives from epistemology and philosophy of science* (pp. 37–49). Routledge, Taylor and Francis Group.

Stuart, M. (2017). How thought experiments increase understanding. In Stuart, M. T., Fehige, Y., and Brown, J. R. (eds.). *The Routledge companion to thought experiments* (pp. 526–44). Routledge.

Stueber, K. R. (2012). Understanding versus explanation? How to think about the distinction between the human and the natural sciences. *Inquiry*, *55*(1), 173–2.

(2017). Empathy and understanding reasons. In Maibom, H. (ed.). *The Routledge handbook of philosophy of empathy* (pp. 137–47). Routledge.

Suchman, A. L., Markakis, K., Beckman, H. B., and Frankel, R. (1997). A model of empathic communication in the medical interview. *Jama*, *277*(8), 678–82.

Symons, X. (2019). Pellegrino, MacIntyre, and the internal morality of clinical medicine. *Theor Med Bioeth*, *40*(3), 243–51.

Tallis, R. (2004). *Hippocratic oaths: Medicine and its discontents*. Atlantic Books.

Taylor, C. (1971). Interpretation and the sciences of man. *Rev Metaphys*, *25*(1), 3–51.

Thagard, P. (2003). Pathways to biomedical discovery. *Philos Sci*, *70*, 235–54.

(2005). What is a medical theory? *Stud Multidiscip*, *3*, 47–62.

Thomasson, A. (2015). What can philosophy really do? *Philos Mag*, (71), 17–23.

Thomasson, A. L. (2017). Metaphysics and conceptual negotiation. *Philos Iss*, *27*(1), 364–82.

Todres, L., Galvin, K. T., and Holloway, I. (2009). The humanization of healthcare: A value framework for qualitative research. *Int J Qual Stud Health Well-being*, *4*(2), 68–77.

Topol, E. (2019). *Deep medicine: How artificial intelligence can make healthcare human again*. Hachette, UK.

Trevarthen, C. (1979). Communication and cooperation in early infancy: A description of primary intersubjectivity. In Bulowa, M. (ed.). *Before speech: The beginning of human communication* (pp. 321–47). Cambridge University Press.

Tridandapani, S., Bhatti, P., and Wick, C. (2019). Patient photographs: Privacy versus protection. *AJR Am J Roentgenol*, *212*(2), 320.

Trout, J. D. (2002). Scientific explanation and the sense of understanding. *Philos Sci*, *69*(2), 212–33.

Turner, Y., and Hadas-Halpern, I. (2008). The effects of including a patient's photograph to the radiographic examination. Radiological Society of North America 2008 Scientific Assembly and Annual Meeting, Chicago, IL.

Ukraintseva, S., Yashin, A. I., and Arbeev, K. G. (2016). Resilience versus robustness in aging. *J Gerontol A Biol Sci Med Sci*, *71*(11), 1533–4.

Van den Berghe, P. L. (1967). *Race and racism: A comparative perspective.* John Wiley and Sons.
Van Inwagen, P. (2008). How to think about the problem of free will. *J Ethics, 12*(3/4), 327–41.
Varelius, J. (2005). Health and autonomy. *Med Health Care Philos, 8*(2), 221–30.
Varga, S. (2015). *Naturalism, interpretation, and mental disorder.* Oxford University Press.
(2019). *Scaffolded minds.* MIT Press.
(2020) Epistemic authority, philosophical explication, and the bio-statistical theory of disease. *Erkenntnis, 85*, 937–56.
Veatch, R. M. (2001). The impossibility of a morality internal to medicine. *J Med Philos, 26*(6), 621–42.
Veit, W. (2021). Experimental philosophy of medicine and the concepts of health and disease. *Theor Med Bioeth, 42*(3), 169–86.
Venkatapuram, S. (2013). Health, vital goals, and central human capabilities. *Bioethics, 27*(5), 271–9.
Walker, C. M., and Gopnik, A. (2013). Causality and imagination. In Taylor, M. (ed.). *The Oxford handbook of the development of imagination* (pp. 342–58). Oxford University Press.
Wartofsky, K., and Zaner, R. M. (1980). Editorial: Understanding and explanation in medicine. *J Med Philos, 5*, 23.
Waskan, J. (2011). Mechanistic explanation at the limit. *Synthese, 183*(3), 389–408.
Weatherall, D. (1996). *Science and the quiet art: The role of medical research in health care.* W. W. Norton.
Wilkenfeld, D. A. (2013). Understanding as representation manipulability. *Synthese, 190*(6), 997–1016.
Williamson, T. (2004). Philosophical "intuitions" and scepticism about judgement. *Dialectica, 58*(1), 109–52.
(2018). *Doing philosophy: From common curiosity to logical reasoning.* Oxford University Press.
Williamson, J. (2019). Establishing causal claims in medicine. *Int Stud Philos Sci, 32*(1), 33–61.
Woodward, J. (2003). *Making things happen: A theory of causal explanation.* Oxford University Press.
(2004). Counterfactuals and causal explanation. *Int Stud Philos Sci, 18*(1), 41–72.
(2010). Causation in biology: Stability, specificity, and the choice of levels of explanation. *Biol Philos, 25*(3), 287–318.
(2015). Interventionism and causal exclusion. *Philos Phenomenol Res, 91*(2), 303–47.
Wootton, D. (2006). *Bad medicine: Doctors doing harm since Hippocrates.* Oxford University Press.
Wright, G. H. (1963). *Norm and action: A logical inquiry.* Routledge and Kegan Paul.

Wright, H. G. (2007). *Means, ends and medical care*. Springer.
Yanikkerem, E., Ay, S., and Piro, N. (2013). Planned and unplanned pregnancy: Effects on health practice and depression during pregnancy. *J Obstet Gynaecol Res, 39*(1), 180–7.
Ylikoski, P. (2013). Causal and constitutive explanation compared. *Erkenntnis, 78*, 277–97.
Zola, I. K. (1972) Medicine as an institution of social control. *Sociol Rev, 20*, 487–504.

Index

Printed in the United States
by Baker & Taylor Publisher Services